Deep Learning for Autonomous Vehicle Control

Algorithms, State-of-the-Art, and Future Prospects

Synthesis Lectures on Advances in Automotive Technology

Editor
Amir Khajepour, *University of Waterloo*

The automotive industry has entered a transformational period that will see an unprecedented evolution in the technological capabilities of vehicles. Significant advances in new manufacturing techniques, low-cost sensors, high processing power, and ubiquitous real-time access to information mean that vehicles are rapidly changing and growing in complexity. These new technologies—including the inevitable evolution toward autonomous vehicles—will ultimately deliver substantial benefits to drivers, passengers, and the environment. Synthesis Lectures on Advances in Automotive Technology Series is intended to introduce such new transformational technologies in the automotive industry to its readers.

Deep Learning for Autonomous Vehicle Control: Algorithms, State-of-the-Art, and Future Prospects
Sampo Kuutti, Saber Fallah, Richard Bowden, and Phil Barber
2019

Narrow Tilting Vehicles: Mechanism, Dynamics, and Control
Chen Tang and Amir Khajepour
2019

Dynamic Stability and Control of Tripped and Untripped Vehicle Rollover
Zhilin Jin, Bin Li, and Jungxuan Li
2019

Real-Time Road Profile Identification and Monitoring: Theory and Application
Yechen Qin, Hong Wang, Yanjun Huang, and Xiaolin Tang
2019

Noise and Torsional Vibration Analysis of Hybrid Vehicles
Xiaolin Tang, Yanjun Huang, Hong Wang, and Yechen Qin
2018

Smart Charging and Anti-Idling Systems
Yanjun Huang, Soheil Mohagheghi Fard, Milad Khazraee, Hong Wang, and Amir Khajepour
2018

Design and Avanced Robust Chassis Dynamics Control for X-by-Wire Unmanned Ground Vehicle
Jun Ni, Jibin Hu, and Changle Xiang
2018

Electrification of Heavy-Duty Construction Vehicles
Hong Wang, Yanjun Huang, Amir Khajepour, and Chuan Hu
2017

Vehicle Suspension System Technology and Design
Avesta Goodarzi and Amir Khajepour
2017

Deep Learning for Autonomous Vehicle Control: Algorithms, State-of-the-Art, and Future Prospects
Sampo Kuutti, Saber Fallah, Richard Bowden, and Phil Barber

ISBN: 978-3-031-00374-5   paperback
ISBN: 978-3-031-01502-1   ebook
ISBN: 978-3-031-00007-2   hardcover

DOI 10.1007/978-3-031-01502-1

A Publication in the Springer series
SYNTHESIS LECTURES ON ADVANCES IN AUTOMOTIVE TECHNOLOGY

Lecture #8
Series Editor: Amir Khajepour, *University of Waterloo*
Series ISSN
Print 2576-8107   Electronic 2576-8131

Deep Learning for Autonomous Vehicle Control

Algorithms, State-of-the-Art, and Future Prospects

Sampo Kuutti
University of Surrey, UK

Saber Fallah
University of Surrey, UK

Richard Bowden
University of Surrey, UK

Phil Barber
Jaguar Land Rover

SYNTHESIS LECTURES ON ADVANCES IN AUTOMOTIVE TECHNOLOGY #8

ABSTRACT

The next generation of autonomous vehicles will provide major improvements in traffic flow, fuel efficiency, and vehicle safety. Several challenges currently prevent the deployment of autonomous vehicles, one aspect of which is robust and adaptable vehicle control. Designing a controller for autonomous vehicles capable of providing adequate performance in all driving scenarios is challenging due to the highly complex environment and inability to test the system in the wide variety of scenarios which it may encounter after deployment. However, deep learning methods have shown great promise in not only providing excellent performance for complex and non-linear control problems, but also in generalizing previously learned rules to new scenarios. For these reasons, the use of deep neural networks for vehicle control has gained significant interest.

In this book, we introduce relevant deep learning techniques, discuss recent algorithms applied to autonomous vehicle control, identify strengths and limitations of available methods, discuss research challenges in the field, and provide insights into the future trends in this rapidly evolving field.

KEYWORDS

artificial intelligence, machine learning, deep learning, neural networks, computer vision, autonomous vehicles, intelligent transportation systems, advanced driver assistance systems, vehicle control, interpretability, safety validation

Contents

List of Figures

List of Tables

Preface

The introduction of new digital technology into the automotive industry has always raised anxieties. From the concerns over the use of floating point arithmetic in the 1980s, through the scepticism over the adoption of the C-language for safety-related control systems, to the introduction of graphical methods and auto coding of real-time algorithms. We now comfortably accept the implementation of nonlinear rules and behaviors for vehicle control. These non-linear behaviors can be considered the precursors to intelligent algorithms, exemplified by the now mature tracking and distance keeping offered by Adaptive Intelligent Cruise Control systems initially researched in the 1990s. Current developments in Artificial Intelligence continue this progress towards more natural and capable vehicle control systems, but quite rightly need a complete understanding of the complex safety issues involved.

This book aims to present a view of current technological capabilities in autonomous vehicle control through deep learning methods, discussing algorithms and techniques used, shortcomings of current systems, research challenges, and safety perspectives. The target audience of the book includes graduate students interested in autonomous vehicles, automotive engineers and researchers who want to learn how these new tools can be applied to the next generation of intelligent transportation systems, as well as deep learning practitioners who aim to apply their skills to self-driving cars.

To this aim, the book starts with an introduction to autonomous vehicle research and the motivation behind this technology in Chapter 1. Chapter 2 lays the foundation of deep learning concepts, which will be relevant to later chapters. This chapter is also intended as a general introduction to deep learning, for those readers who are less familiar with this field. The review of early concepts and state-of-the-art deep learning algorithms for autonomous vehicle control are given in Chapter 3. The challenges related to safety validation of these black box systems are discussed further in Chapter 4. Finally, conclusions and future trends are outlined in Chapter 5.

The research on which this book is based on was carried out at the Connected and Autonomous Vehicles Lab (CAV-LAB) at University of Surrey, together in partnership with the Centre for Vision, Speech, and Signal Processing (CVSSP) and Department of Computer Science at University of Surrey as well as Jaguar Land Rover. The research project was supported by the UK-EPSRC grant EP/R512217/1 and Jaguar Land Rover.

Sampo Kuutti, Saber Fallah, Richard Bowden, and Phil Barber
January 2019

CHAPTER 1

Introduction

Autonomous vehicles have the potential to transform our transportation systems in terms of safety and efficiency. The steady increase in the number of vehicles on the road has led to increased pressure to solve issues such as traffic congestion, pollution, and road safety. The leading answer to resolving these issues among the research community is self-driving cars [1–3]. For instance, according to the World Health Organization, an estimated 1.3 million people die in road accidents yearly [4]. Meanwhile, up to 90% of all car accidents are estimated to be caused by human errors [5], therefore autonomous vehicles can provide significant safety improvements by eliminating driver errors. Further benefits provided by autonomous vehicles include better fuel economy, reduced pollution, car sharing, increased productivity, and improved traffic flow [6–9].

Autonomous vehicles generally consist of the five functional components shown in Fig. 1.1: Perception, Localization, Planning, Control, and System Management [10]. Perception observes the environment around the vehicle and identifies important objects such as traffic signals and obstacles. Localization maps the surrounding environment and identifies the location of the vehicle in absolute position. Planning uses the input from perception and localization to determine the high-level actions the vehicle will take in terms of routes, lane changes, and desired velocity. Control module oversees carrying out low-level actions indicated by the planning, such as steering, accelerating, and braking for the vehicle. System management oversees the operation of all the modules and provides the Human-Machine Interface.

Among the earliest autonomous vehicles are the projects presented by Carnegie Mellon University for driving in structured environments [11] and University of Bundeswehr Munich for highway driving [12] in the 1980s. Since then, projects such as DARPA Grand Challenges [13, 14] have continued to drive forward research in autonomous vehicles. Besides academic research, car manufacturers and tech companies have also begun developing their own autonomous vehicle capabilities. These early steps toward autonomy have led to multiple Advanced Driver Assistance Systems (ADAS) such as Adaptive Cruise Control (ACC), Lane Keeping Assistance (LKA), and Lane Departure Warning (LDW) technologies, which provide modern vehicles with level 1-2 autonomy (see Fig. 1.2) [15]. While these technologies have increased the safety of modern vehicles and made driving easier, the end goal in autonomous vehicle research is to achieve level 5 autonomy, with a fully autonomous vehicle which does not require human intervention. Therefore, these partially autonomous systems pave the way for the future autonomous vehicles.

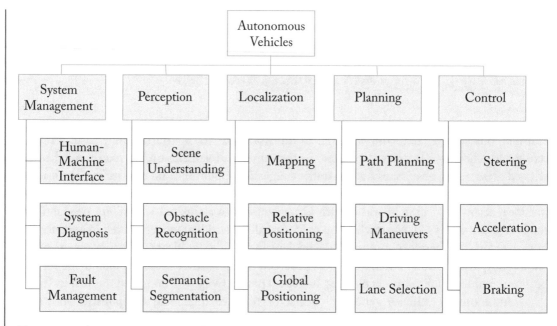

Figure 1.1: Autonomous vehicle functional units.

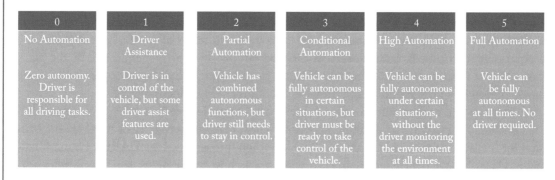

Figure 1.2: The six levels of autonomy in road vehicles, as defined by the Society of Automotive Engineers (SAE).

Early autonomous vehicles used a perception-planning-control framework, with each function achieved separately [16]. Perception relied on accurate sensor data, with multi-sensor setups and sensor fusion used to capture the driving environment accurately. Planning and control were achieved through rule-based systems and classical controllers were often based on linearized (or otherwise simplified) vehicle models [17–19]. The downside of this approach is that the control parameters required hand-tuning based on simulation and field test results, which was very time intensive and resulted in systems which had difficulty generalizing to a wide variety of scenarios [20, 21]. Moreover, linearized vehicle models can cause significant inaccuracies in the highly nonlinear regions of driving, which meant that these control methods were infeasible or did not scale well in certain scenarios [22, 23]. The shortcomings of early autonomous vehicle systems led researchers to look for alternative solutions. Recently, the rise of deep learning has drastically changed research in several areas of Artificial Intelligence (AI), and significantly improved the state-of-the-art in fields such as image classification and speech recognition [24–26]. The powerful function approximation capabilities of Deep Neural Networks (DNNs) has also motivated the use of deep learning in several autonomous vehicle applications, including planning and decision making [27–29], vehicle-to-vehicle communications [30, 31], perception [32–34], as well as mapping and localization [35–37].

Deep learning has gained significant interest as a promising solution to autonomous vehicle control. Deep learning not only provides excellent performance in control applications but can also provide the capability to generalize its previously learned rules to new scenarios [38–42]. Rather than requiring a formal specification of the exact behavior as you would in a rule-based system, deep learning enables the system to learn a general behavior from examples either via demonstration or interaction with the environment. The powerful representational power of DNNs and generalization capability given by deep learning makes these systems well suited for complex and dynamic tasks and operational environments [43–45], such as those seen by an autonomous vehicle. However, the disadvantage of deep learning is its complex and opaque nature. This poses a challenge for the safety validation and verification of deep learning algorithms. Due to the lack of interpretability in terms of what the neural network has learned and how it makes its decisions, there are currently no known methods for guaranteeing the safety of such a system to a certifiable level. Therefore, new methods for validating the safety of deep learning systems in autonomous vehicles are required.

This book offers a comprehensive view of the current state of autonomous vehicle control techniques through deep learning. A sampling of the most recent research works is presented in this field, and different approaches are compared and analysed. Relevant research challenges and future research directions are also discussed. The contents of the remainder of the book are as follows. Chapter 2 gives the reader a brief overview of deep learning theory. Although a comprehensive description of the whole deep learning field is out of the scope of this book, the chapter serves as a good introduction to deep learning for beginners in this field, and provides a summary of key deep learning concepts discussed later in the book. Further reading for a more comprehensive look into deep learning concepts are also recommended. Chapter 3 reviews current work done in deep learning vehicle control, giving the reader a comprehensive view of the state-of-the-art approaches in the field. The strengths and limitations of different approaches are described through comparative analysis. Furthermore, research challenges and future research directions are discussed. Chapter 4 discusses the safety validation of deep neural networks in autonomous vehicles. The difficulty of validating the safety of these opaque systems is discussed and potential validation and verification techniques are introduced. Finally, concluding remarks are given in Chapter 5.

CHAPTER 2

Deep Learning

Machine learning is a powerful tool in AI, where rather than engineer an algorithm to complete a task, the designer can provide the algorithm with examples (i.e., training data) and the algorithm learns to solve the task on its own. Given enough training data, machine learning algorithms can optimize their solution to outperform traditional programming methods. Artificial neural networks are a promising tool for machine learning methods, and have gained significant attention since the discovery of techniques for training larger multilayer neural networks, called deep neural networks. This class of techniques utilizing deep neural networks for machine learning are referred to as deep learning. Deep learning has been developing rapidly in recent years and has shown great promise in fields such as computer vision [24], speech recognition [25], and language processing [26]. The aim of this chapter is to provide the reader with a brief background on neural networks and deep learning methods which are discussed in the later sections.

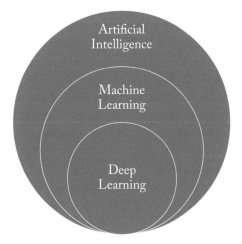

Figure 2.1: Deep learning is a subset of machine learning techniques.

2.1 NEURAL NETWORK ARCHITECTURES

The structure of artificial neural networks is inspired by the neural structure of the human brain, with layers of interconnected neurons transferring information. Figure 2.2 illustrates a simple

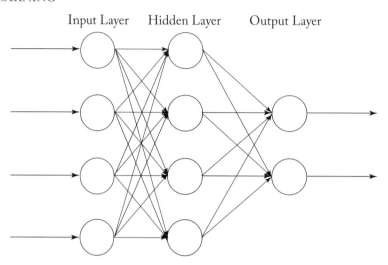

Figure 2.2: Feedforward neural network.

feedforward neural network. The first layer (on the left) is the input layer, with each neuron representing an input to the neural network. The second layer is the hidden layer. Although in the neural network shown in Fig. 2.2 there is only one hidden layer, this number can vary. The final layer is the output layer, with each neuron representing an output from the neural network. The neurons send their values to neurons in the next layer. Typically, each neuron is connected to all neurons in the next layer. This is called a fully connected neural network. Each connection also has a weight, which multiplies the input to the neuron in the next layer. Therefore, the input at the next layer is a weighted summation of the outputs of the previous layer. The weights of these connections are changed over time as the neural network learns. The neural network tunes these weights over the training process to optimize its performance at a given task. Generally, a loss function is used as a measure of the network error, and network weights are updated such that the loss is minimized during training.

The hidden neurons also normalize their output by applying an activation function to its input, such as a Rectified Linear Unit (ReLU) or a sigmoid function. The activation function is helpful as it introduces some nonlinearity in the neural network. Consider a case of a neural network with no nonlinearities, since the output of each layer is a linear function, and a sum of linear functions is still a linear function, the relationship between the input and the network output could be described by the function $F(x) = mx$. To update the weights using gradient descent, the gradient of this function would be calculated as m. Therefore the gradient is not a function of the inputs x. Moreover, in the case of linear activations, there is no benefit in making the network deeper as the function $F(x) = mx$ can always be estimated by a neural network with a single hidden layer. However, by using nonlinear activation functions, we can make the networks deeper by introducing more hidden layers and enabling the networks to model more complex relationships. A set of nonlinear activation functions commonly used in deep neural networks is illustrated in Fig. 2.3.

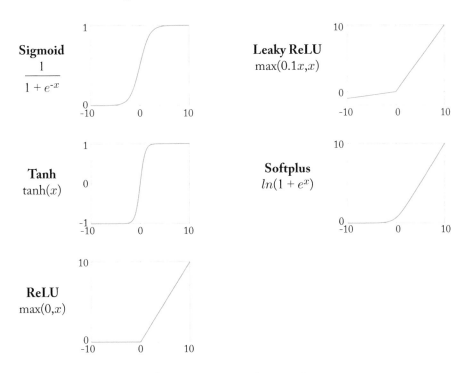

Figure 2.3: Common activation functions in neural networks.

The second class of neural network architectures commonly used are the Recurrent Neural Networks (RNNs). RNNs provide the neural network with temporal context. As shown in Fig. 2.4, they are essentially feedforward neural networks, but with the outputs of neurons from the previous time step as an additional input into the network. This allows the network to remember previous inputs and outputs, providing it with memory of previous states and actions. This is useful in domains where the inputs and outputs are dependent on the inputs and outputs of previous time steps. The downside of RNNs is the increased difficulty in training the network [46–50]. Commonly used recurrent networks include the Long Short-Term Memory (LSTM) [51] and Gated Recurrent Unit (GRU) [52].

Convolutional neural networks (CNNs) are used for tasks where spatial context is required, such as image processing tasks. The main types of layers utilized in CNNs are the input, convolution, pooling, ReLU, and fully connected layers. A typical CNN structure is shown in Fig. 2.5. The input layer is typically a three-dimensional input, consisting of the raw pixel values of the image and the RGB color channel values. The convolutional layers are used to compute the dot product of the convolution filter and a local region of the previous layer. The convolution

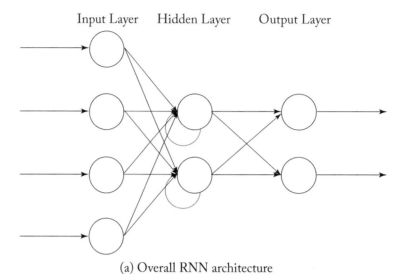

(a) Overall RNN architecture

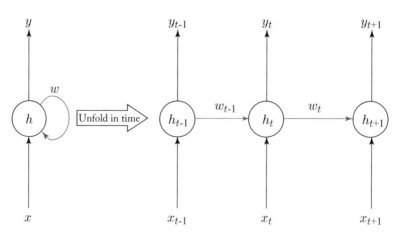

(b) A hidden neuron in an RNN unfolded in time

Figure 2.4: Recurrent neural network, where the red arrows represent temporal connections.

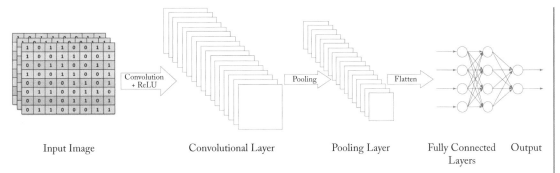

Figure 2.5: Convolutional neural network.

filter produces an activation map, representing the responses of the filter at each region of the input. As the network is trained, the filters learn to recognize useful features such as edges and corners, or even more specific features such as faces or wheels at the later layers of the network. Pooling layers reduce the spatial size of the previous layer outputs to reduce the number of parameters in the network and the computation required. The most common pooling strategy is max pooling, which applies a max filter to regions of the initial representation to provide an abstracted form of the representation. The ReLU layers apply a nonlinear activation function, similar to activation functions in other neural networks. Finally, the fully connected layer is simply a fully connected feedforward layer at the end of the network used to perform tasks such as classification or regression [24, 53].

2.2   SUPERVISED LEARNING

Having chosen a neural network architecture, the second choice when tackling a deep learning problem is the choice of learning strategy. The learning strategy dictates how the neural network's weights are updated in response to training examples, allowing the neural network to learn trends in the data and identify rules that hopefully generalize to new data. There are multiple choices for learning strategies. However, most deep learning algorithms discussed in later sections can be broadly classified as supervised or reinforcement learning, therefore these two learning strategies will be described here.

Supervised learning utilizes labeled data to learn rules about the problem domain. Labeled data here means that the algorithm is given examples where the input as well as the correct output for each input are known. For example, in an image classification problem, the labeled data would be images labeled with the correct classification. Therefore, the neural network can make its own prediction and compare it to the correct answer. The difference between the correct output and the network output is referred to as the error which is used to update the network weights. After training, the network will learn to choose the correct output for input data without the labels.

Supervised learning benefits from a relatively fast training process. However, supervised learning requires large amounts of data to achieve adequate performance and generalization capability. Moreover, labeled data may not be available beyond the initial training process which limits its application to some domains.

The network weight update is typically calculated using Stochastic Gradient Descent (SGD) or other variants based on it. SGD optimizes the network parameters, θ, such that the loss function, $\mathcal{L}$, is minimized over the training set $x_{k=1...N}$

$$\theta = \underset{\theta}{\operatorname{argmin}} \frac{1}{N} \sum_{k=1}^{N} (\mathcal{L}(x_k, \theta)). \tag{2.1}$$

This is typically done in mini-batches of size M, to provide an estimate for the loss function over the training set. Smaller batches as an approximation for weight updates are required as the training sets are generally too large to be used for computing the loss function. On the other hand, it is advantageous to use these mini-batches over computing the loss function for each example for two reasons. First, a larger sample size gives us a better estimation of the true loss function over the training set [54]. Second, the parallel computing capabilities offered by modern Graphics Processing Units (GPU) enable significant speed benefits when computing the loss function over multiple training examples. A further hyperparameter, the learning rate η, is used to control the size of the weight updates during training. For instance, if we were to use mean error between predicted outputs, y, and labeled outputs, $\hat{y}$, as the loss function, the loss function and weight updates would then be given by

$$\mathcal{L} = \frac{1}{M} \sum_{k=1}^{M} (\hat{y}_k - y_k) \tag{2.2}$$

$$\theta = -\eta \frac{\partial \mathcal{L}}{\partial \theta} = -\frac{\eta}{M} \sum_{k=1}^{M} (\hat{y}_k - y_k) \left(-\frac{\partial y_k}{\partial \theta} \right), \tag{2.3}$$

where $\mathcal{L}$ is the loss, M is the mini-batch size, y_k is the output for the kth example, $\hat{y}$ is the desired output, θ are the network parameters, and η is the learning rate.

2.3   REINFORCEMENT LEARNING

On the other hand, reinforcement learning enables an agent to learn through interactions with the environment and is often utilized to control dynamic systems. Reinforcement learning is typically modeled as a Markov Decision Process (MDP) which can be formally described as a tuple $(\mathcal{S}, \mathcal{A}, \mathcal{P}, \mathcal{R})$, where $\mathcal{S}$ denotes the state space, $\mathcal{A}$ represents the action space, $\mathcal{P}$ denotes the state transition probability model, and $\mathcal{R}$ represents the reward function. The state space $\mathcal{S}$ is a set of states which contain all the features of the local environment required to predict future states and measure the performance of the agent. The action space $\mathcal{A}$ is a set of actions available to the agent in any given state. The state transition probability model $\mathcal{P}$ describes the probability of reaching a certain state by performing a given action in the current state. However, some reinforcement learning algorithms choose not to model the state transition probabilities, referred to as model-free methods [55]. The MDP is typically episodic, where the state is reset to a starting state s_0 after a set time T or based on a termination condition, such as a successful completion of the task. In reinforcement learning, an agent interacts with an environment at each time step and aims to learn from its own actions. The general reinforcement learning process is summarized in Fig. 2.6. At each time-step t, the agent observes a set of states s_t and then takes an action a_t from a possible set of actions $\mathcal{A}$ according to its policy $\pi(s_t)$. The environment then transitions based on the action a_t, and the agent observes the next set of states s_{t+1} and receives a reward r_t according to the reward function $\mathcal{R}$. The aim of the agent is to maximize the total accumulated return R_t [56–58].

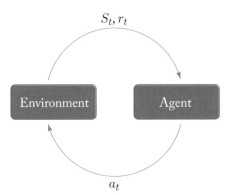

Figure 2.6: Reinforcement learning process. Agent observes states s_t, takes action a_t, receivers reward r_t, and new observation s_{t+1}. The transition $\{s_t, a_t, r_t, s_{t+1}\}$ is then used to update the network parameters.

The reward can be either sparse or dense. With a sparse reward function, the agent receives rewards only following certain events. A common example is a binary reward, where the agent only receives a reward upon successful completion of the task. The advantage of this approach is its simplicity; it is easy to define a success (or a failure) for most tasks, and the agent learns which state it should end up in. The problem, however, is that the majority of time the agent is receiving no rewards, and therefore no indication on how useful its actions were. Thus, sparse reward functions may require a significant amount of exploration before the agent completes the task successfully. On the other hand, in a dense reward structure, the agent is given a reward often (e.g., every time step), providing a continuous feedback of the agent's performance. This means that the agent receives an estimate of how useful each action was in their respective states.

Reinforcement learning algorithms can generally be divided into three classes [59]: value-based, policy gradient, and actor-critic methods. Value-based methods (e.g., Q-learning [60–62] or SARSA [63]) estimate a state-action value function (also called the quality or Q function) $Q(s_t, a_t)$, which maps each possible action to a value for each state:

$$Q(s_t, a_t) = \mathbb{E}[R_t | s_t = s, a].\tag{2.4}$$

The resulting policy chooses the action in a greedy fashion, maximizing the state-action value function as

$$\pi(s_t) = \underset{a}{\mathrm{argmax}}\, Q(s_t, a_t).\tag{2.5}$$

The downside of value-based methods is that there is no guarantee on the optimality of the resulting policy [64–67].

On the other hand, policy gradient methods (e.g., REINFORCE [68–70]) do not estimate the value function. Instead, policy gradient methods estimate the optimal policy by parametrizing the policy $\pi(s)$ by the parameter vector θ, and calculating the gradient of the cost function J with respect to θ as

$$\nabla_\theta J = \frac{\partial J}{\partial \pi_\theta} \frac{\partial \pi_\theta}{\partial \theta}\tag{2.6}$$

and the parameters are then updated in the direction of the gradient. The advantage of the policy gradient approach over estimating the value function, is improved convergence. However, the disadvantage is high variance in the estimated policy gradient [71, 72].

The third class of algorithms are actor-critic methods (e.g., A3C [73] or DDPG [74]), which are hybrid methods combining both value-based and policy gradient methods, in order to exploit the advantages of both techniques [64, 75, 76]. Actor-critic algorithms use two networks to choose actions for a given state. The critic network estimates how valuable being in a given state is, by estimating the expected return at current state s_t under current policy π with the value function $V(s)$:

$$V(s) = \mathbb{E}[R_t | s_t = s, \pi].\tag{2.7}$$

Meanwhile, the actor network estimates the optimal policy function $\pi^*(s_t)$, which is the policy that maximizes the expected return. Therefore, the actor network maps observed states to actions, while the critic network evaluates the actions taken. Estimating the value function is useful, since we cannot know the actual value of the actions taken (i.e., total discounted rewards for an episode) until the episode has finished. The critic network allows the algorithm to estimate the value of the actions taken during training. Therefore, unlike policy gradient methods where total rewards are calculated at the end of the episode so that network weights can be updated, estimating the value function lets us update the network weights before the episode is finished, such that we can update the weights multiple times each episode [73]. The actor-critic framework is illustrated in Fig. 2.7. During training, the agent interacts with the environment,

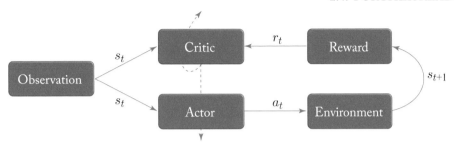

Figure 2.7: An actor-critic agent interacting with the environment. Dashed lines represent a parameter update.

and based on the actions chosen by the actor network, the agent is given a reward. The value function estimated by the critic network is then used to update the parameters of both networks.

2.4   FURTHER READING

In this chapter, a brief overview of relevant deep learning concepts was given. After reading this chapter, you should have a basic understanding of how deep learning works, and the different general approaches one can use to tackle deep learning problems. However, this chapter only touched on the concepts relevant to later sections of the book, since a full review of deep learning theory is out of the scope of this book. For a broader and more in-depth view of deep learning theory, we provide the interested reader with some useful books in this field. For beginners, *Neural Networks and Deep Learning* by Nielsen [77] provides an intuitive introduction to neural networks. For more in-depth reading, *Deep Learning* by Goodfellow et al. [78] provides a comprehensive view of deep learning theory. For an introduction to reinforcement learning, we recommend *Reinforcement Learning: An Introduction* by Sutton and Barto [56]. For a deeper look at the mathematical background in machine learning, we recommend the books by Deisenroth et al. [79] and Hastie et al. [80]. For a more hands-on introduction to deep learning, the books by Chollet [81] and Géron [82] provide useful examples with code for those looking to learn how to implement deep learning algorithms.

CHAPTER 3

Deep Learning for Vehicle Control

Deep learning has provided a promising solution to autonomous vehicle control. The strong function approximation for learning nonlinear functions, generalization capability offered through learning from big data, and highly scalable properties to high-dimensional observation-action mappings enables deep learning to outperform hand-engineered control techniques. For these reasons, there has been several approaches to using deep learning to control autonomous vehicles, with a variety of sensor configurations, control objectives, network architectures, and learning algorithms. Thus, in this chapter we review existing methods in the literature, analyze the strengths and limitations of different techniques, discuss main research challenges, and make recommendations for the direction of future research. Since multiple research projects have focussed on learning a single driving action, the discussion on control techniques in this chapter is broken into three sections: lateral (steering), longitudinal (acceleration and braking), and full vehicle control.

3.1 AUTONOMOUS VEHICLE CONTROL

3.1.1 LATERAL CONTROL

The earliest example of autonomous vehicle control with neural networks is the Autonomous Land Vehicle In Neural Network (ALVINN) by Pomerleau in 1989 [83, 84]. ALVINN used a single hidden layer feedforward network to predict a steering angle based on an image input. The output of the network was discretized into 30 different possible steering angle values. The training used supervised learning where the training data consisted of images from a front-view camera labeled with the steering commands of the human driver. To improve the robustness of the system and allow the network to correct from mistakes not seen in the training examples, the author augmented the data set with additional synthesized data; images in the training set were shifted and rotated to simulate the vehicle in different parts of the road laterally. In order to avoid bias toward recent inputs in the network a buffering solution was developed to store previously encountered steering patterns, with up to four patterns stored at any time. This buffer was updated periodically to ensure no right or left bias existed in the stored data. The use of data augmentation and a training buffer were shown to significantly improve the system performance. The trained network was tested by driving a vehicle at speeds ranging from 5–55 mph, achieving steering without interventions for distances of up to 22 miles. The average distance from the center of the lane during testing was 1.6 cm compared to that of 4.0 cm under human control. This demonstrated that neural networks can learn to steer a vehicle from expert demonstration through supervised learning. In the first work to train a network for lateral control through reinforcement learning, Yu [85] extended the approach of ALVINN. The proposed approach trained a feedforward network to steer a vehicle from images, using reinforcement learning so that the network could continue to learn from its experiences even after deployment.

Since these early works, the advancement in computational capabilities through Graphics Processing Units (GPU) has enabled researchers to use deeper models with significantly increased amounts of training data. Using CNNs to steer a sub-scale vehicle, Muller [86] demonstrated the capabilities of deeper models for autonomous vehicle control. The vehicle, Darpa Autonomous VEhicle (DAVE), learned to drive off-road with speeds up to 2 m/s while avoiding obstacles in its path. Building on the approach of DAVE, NVIDIA's end-to-end steering system, PilotNet, was trained to keep a vehicle in its lane [87]. PilotNet was trained on images captured from a front-facing camera as a human driver steered the vehicle. The collected training set was used to train a CNN to predict steering angles based on an image input. To ensure the network could recover from mistakes, the data set was augmented with random shifts and rotations from three cameras (left, center, right) to simulate the vehicle veering away from its lane, without having to actually drive the vehicle off the road. The data set was captured in a variety of weather and lighting conditions, adding up to 72 hours of collected data. The trained network was initially tested in a simulation environment, and following positive results from the simulator, the network was tested in the real world. Initial on-road testing demonstrated the network can keep the vehicle in its lane, with up to 98% autonomy, and highway testing demonstrated capability to drive up to 10 miles without human interception.

A similar approach was used by Rausch et al. [45] to train a CNN for steering in simulation. The network was trained on 15 minutes of collected driving data, with a human steering the vehicle within the simulation. The trained network was evaluated in the simulation and demonstrated it had learned a good estimation of the demonstrator's driving policy even with minimal training data. However, the steering signal was shown to exhibit oscillatory behavior. This can be explained by the fact that the system estimates the required steering angle at each frame, with no context regarding the previous states or actions. This means that as the system progresses to the next time step, the system estimates a steering output without considering the previous action. This can result in the steering signal between two time steps to vary significantly, leading to large changes in the steering wheel angle within short time periods. A potential solution to this issue would be to use a RNN to provide temporal context in the neural network. For example, introducing temporal context into the model, Eraqi et al. [88] trained a CNN-RNN model to predict steering values from images. The proposed model was trained and validated on the publicly available comma.ai data set. Although no live testing (where the network would have to control the vehicle and correct any mistakes it had made) was used, the trained network was shown to learn steering policies from images that compared favorably to simple CNN models without temporal dependencies. Therefore, the use of recurrent architectures can provide the vehicle with a smooth and consistent driving policy.

A summary of the research works covered in this section can be seen in Table 3.1. The advancements in parallelizable computing and available data sets have enabled the field to move to deeper models trained on increased amounts of data. The majority of lateral control systems have chosen to use supervised learning, where the training data is provided with a front-facing camera and a human steering the vehicle. Although many of these systems were trained for simplified tasks (e.g., lane keeping only) and/or in simplified training environments (e.g., simulation/no adjacent vehicles) the results demonstrate that deep learning can be used to train systems for mapping high-dimensional images straight to steering signals. Thus, important developments

Table 3.1: A comparison of lateral control techniques

Ref.	Learning Strategy	Network	Inputs	Outputs	Pros	Cons
[83], [84]	Supervised learning	Feedforward network with one hidden layer	Camera image	Discretized steering angles	First promising results for neural network-based vehicle controllers	Simple network and discretized outputs degrade performance
[85]	Reinforcement Learning	Feedforward network with one hidden layer	Camera image	Discretized steering angles	Supports online learning	Simple network and discretized outputs degrade performance
[86]	Supervised learning	six-layer CNN	Camera image	Steering angle	Robust to environmental diversity	Large errors, trained and tested on a sub-scale vehicle model
[87]	Supervised learning	nine-layer CNN	Camera image	Steering angle values	High level of autonomy during field test	Only considers lane following, occasionally requires interventions by the driver
[45]	Supervised learning	eight-layer CNN	Simulated camera image	Steering angle values	Learns from minimal training data	Oscillatory behavior of the steering signal
[88]	Supervised learning	CNN-RNN	Camera image	Steering angle values	Considers temporal dependencies	Instability of RNN training, No live testing

have been made toward high-level autonomy and we can expect further advancements using DNNs to drive vehicles in more complex scenarios in the future.

3.1.2   LONGITUDINAL CONTROL

Deep learning also offers multiple advantages for longitudinal vehicle control. The longitudinal control of an autonomous vehicle can be described as an optimal tracking control problem for a complex nonlinear system [89, 90], and is therefore poorly suited to control systems based on linearized vehicle models or other simplified analytical solutions [91]. Traditional control systems used in longitudinal ADAS systems, such as Adaptive Cruise Control, provide the autonomous vehicle with poor adaptability to different scenarios. However, deep learning solutions have shown strong performance in nonlinear control problems and can learn to perform a task without knowledge of the system model [43, 44, 92, 93].

In early works of neural longitudinal control, Dai et al. [94] presented a fuzzy reinforcement learning control algorithm for longitudinal vehicle control. The proposed algorithm combines reinforcement learning with fuzzy logic, where the reinforcement learning was based on Q-learning and fuzzy logic used a Takagi-Sugeno-type fuzzy inference system. The Q-learning module estimates the optimal control action in the current state, while the fuzzy inference system produces the final control output based on the estimated action value. The reward function for Q-learning was set up based on the distance between the host vehicle and lead vehicle, to encourage the vehicle to follow at a safe distance. The trained system was evaluated in a simulated car following scenario, and the vehicle was shown to successfully drive the vehicle without failures after 68 trials. However, while this approach demonstrated reinforcement learning can be used to successfully follow vehicles in front, a reward function based on a single objective can lead to unexpected behavior. The reward function is the method by which the designer can signal the desired behavior to the reinforcement learning agent, and should therefore accurately represent the control objective. Therefore, reward functions for autonomous vehicles should encourage safe, comfortable, and efficient driving strategies. To achieve this, multi-objective reward functions should be investigated.

For instance, Desjardins and Chaib-Draa [23] used a multi-objective reward function based on time headway and time headway derivative. This encouraged the agent to maintain an ideal time headway to the lead vehicle (set to 2 s in experiments), but the time headway derivative term in the reward function also rewarded actions which moved the vehicle toward the ideal time headway and penalized actions which moved it farther away from the ideal time headway. The proposed approach was implemented in a cooperative adaptive cruise control system using a policy gradient algorithm. The chosen network architecture was a feedforward network, with a single hidden layer consisting of 20 neurons and an output layer with 3 outputs (accelerate, brake, do nothing). Ten training runs were completed, and the best performing network was chosen for testing. During testing, adequate vehicle following behavior was demonstrated, with time headway errors of 0.039 s achieved in emergency braking scenarios. However, the downside was that oscillatory velocity profiles were observed, which poses safety and passenger comfort issues. Similarly to oscillatory outputs in lateral control, this could potentially be resolved with the use of RNNs. Another potential solutions would be to design a reward function with an additional term for drive smoothness. Such a reward function was used, for instance, by Huang et al. [95], who used an actor-critic algorithm for autonomous longitudinal control. The multi-objective reward function considered the velocity tracking error and drive smoothness. This helps ensure that no sharp accelerations or decelerations are used unnecessarily. This results in a control policy which is comfortable for occupants in the vehicle. However, no adjacent vehicles or safety were considered in this work.

An algorithm for a personalized ACC system was presented by Chen et al. [96]. The proposed approach used Q-learning to estimate the desired vehicle velocity at each time-step, which was then achieved using a Proportional-Integral-Derivative (PID) controller. A single hidden layer feedforward network was used to estimate the Q-function. The system was evaluated in simulation, based on driving smoothness, passenger comfort, and safety. A similar approach was used by Zhao et al. [97], who used an actor-critic algorithm to learn personalized driving styles for an ACC system. The reward function considered driver habits, passenger comfort, and vehicle safety. The trained network was tested in a hardware-in-the-loop simulation and compared to PID and Linear Quadratic Regulator (LQR) controllers. The proposed algorithm was shown to outperform the traditional PID and LQR controllers in the test scenarios, demonstrating the power of reinforcement learning for longitudinal control.

A collision avoidance system using reinforcement learning was developed by Chae et al. [98]. The proposed approach used a Deep Q-Network (DQN) algorithm to choose from four different discrete deceleration values. A two-term reward function was used, which considered collision avoidance and avoiding high risk situations. The reward function was balanced to avoid too conservative or reckless braking policies using these conflicting objectives. A replay memory was used to improve convergence of training and an additional "trauma memory" of collisions was used as well. The trauma memory improved the stability of learning, by ensuring the agent learns to avoid collisions since these events were very rare during training and a random sampling from the standard replay memory would therefore rarely include them in the parameter update. The system was trained to avoid collisions with pedestrians, which entered in front of the vehicle at different distances. During evaluation, the collision avoidance was tested for different Time-To-Collision (TTC) values, with 10,000 tests for each value. For TTC values above 1.5 s, collisions were avoided with 100% success, while at the lowest TTC value of 0.9 s the collision rate was 61.29%. The system was also tested in the Euro NCAP standard testing procedure (CVFA and CVNA tests [99]) and the system passed both tests successfully. Therefore, the system was deemed to provide adequate collision avoidance for autonomous vehicles.

The use of reinforcement learning has been favored for longitudinal control and has yielded significant improvements in longitudinal control for autonomous vehicles. The use of one-dimensional measurements, such as intervehicular distance, as inputs means that the network complexity can be reduced when compared to those seen in lateral control. This has meant that instead of vision based set-ups, ranging sensors have been more widely used. Moreover, these one-dimensional measurements also enable easier design of reward functions for reinforcement learning since they often directly serve as metrics for performance. While reinforcement learning has provided good results, it has multiple drawbacks. The major drawback of reinforcement learning is the large amount of experiences required for training convergence, which can be expensive and/or time consuming to collect [56, 100]. In contrast, supervised learning methods often lack the level of generalization offered by reinforcement learning, but tend to provide faster convergence, given a suitably high quality data set can be obtained. For these reasons, combining the two learning methods can provide significant advantages. Using supervised learning as a pre-training stage for reinforcement learning, Zhao et al. [22, 101, 102] presented their Supervised Actor-Critic (SAC) algorithm for longitudinal control. The system was trained for vehicle following on dry road conditions, and then evaluated on both dry and wet road conditions. During evaluation, the SAC controller was compared to a PID and a supervised learning (without reinforcement learning) controller and the SAC controller was found to exhibit the best performance. These results show that combining supervised and reinforcement learning can leverage the advantages of both strategies.

A summary of research works discussed in this section can be seen in Table 3.2. The general trend in longitudinal control has been to utilize the generalization capabilities of reinforcement learning to perform well in different environments. Unlike lateral control, the inputs seen here are mostly lower-dimensional measurements of host vehicle states and measurements relative to nearby vehicles (or other objects), which can be obtained from ranging sensors. This means that network complexity can be reduced, and there is no need to use CNNs. However, as discussed in this chapter, in reinforcement learning, the reward function needs to successfully represent the desirability of being in any given state. For a complex task such as driving, this will require considering multiple conflicting objectives, such as desired velocity, passenger comfort, driver habits, and safety. A poorly designed reward function can lead to training divergence, poor control performance, or unexpected behavior.

3.1.3 FULL VEHICLE CONTROL

There has also been some recent interest in investigating vehicle controllers which control all actions of the vehicle (steering, acceleration, braking) simultaneously. For instance, Zheng et al. [103] proposed a decision-making system for an autonomous vehicle in a highway scenario based on reinforcement learning. The system considered three factors in the reward function: (1) safety, evaluated based on distances to adjacent vehicles; (2) smoothness, based on accumulation of longitudinal speed changes over time; and (3) velocity tracking, based on the difference between current velocity and desired velocity. The authors used a Least Squares Policy Iteration approach to find the optimal decision-making policy. The system was then evaluated in a simple highway simulation, where the vehicle had to undergo multiple overtake maneuvres. The system

Table 3.2: A comparison of longitudinal control techniques

Ref.	Learning Strategy	Network	Inputs	Outputs	Pros	Cons
[94]	Fuzzy reinforcement learning	Feedforward network with one hidden layer	Relative distance, relative speed, previous control input	Throttle angle, brake torque	Model-free, continuous action values	Single term reward function
[23]	Reinforcement learning	Feedforward network with one hidden layer	Time headway, headway derivative	Accelerate, brake, or no-op	Maintains a safe distance	Oscillatory acceleration behavior, no term for comfort in reward function
[95]	Reinforcement learning	Actor-Critic Network	Velocity, velocity tracking error, acceleration error, expected acceleration	Gas and brake commands	Learns from minimal training data	Noisy behavior of the steering signal
[98]	Reinforcement learning	Feedforward network with five hidden layers	Vehicle velocity, relative position of the pedestrian for past five time steps	Discretized deceleration actions	Reliably avoids collisions	Only considers collision avoidance with pedestrians
[96]	Reinforcement learning	Feedforward network with one hidden layer	Relative distance, relative velocity, relative acceleration (normalized)	Desired acceleration	Provides smooth driving styles, learns personal driving styles	No methods for preventing learning of bad habits from human drivers
[97]	Reinforcement learning	Actor-Critic Network	Relative distance, host velocity, relative velocity, host acceleration	Desired acceleration	Performs well in a variety of scenarios, safety and comfort considered, learns personal driving styles	Adapting unsafe driver habits could degrade safety
[22]	Supervised reinforcement learning	Actor-Critic Network	Relative distance, relative velocity	Desired acceleration	Pre-training by supervised learning accelerates learning process and helps guarantee convergence, performs well in critical scenarios	Requires supervision to converge, driving comfort not considered

could successfully complete all overtaking manoeuvres. In contrast, Shalev-Shwartz et al. [104] consider a more complex scenario in which an autonomous vehicle has to operate around unpredictable vehicles. The aim of the project was to design an agent which can pass a roundabout safely and efficiently. The performance of the agent was evaluated based on (1) keeping a safe distance from other vehicles at all times, (2) the time to finish the route, and (3) smoothness of the acceleration policy. The authors utilized a RNN to accomplish this task, which was chosen due to its ability to learn the function between a chosen action, the current state, and the state at the next time step without explicitly relying on any Markovian assumptions. Moreover, by explicitly expressing the dynamics of the system in a transparent way, prior knowledge could be incorporated into the system more easily. The described method was shown to learn to slowdown when approaching roundabouts, to give way to aggressive drivers, and to safely continue when merging with less aggressive drivers. In the initial testing, the next state was decomposed into a predictable part, including velocities and locations of adjacent vehicles, and a non-predictable part, consisting of the accelerations of the adjacent vehicles. The dynamics of the predictable part of the next state was provided to the agent. However, in the next phase of testing, all state parameters at the next time step were considered unpredictable and instead had to be learned during training. The learning process was more challenging in these conditions, but still succeeded. Additionally, the authors claimed that the described method could be adapted to other driving policies such as lane change decisions, highway exit and merge, negotiation of the right of way in junctions, yielding for pedestrians, and complicated planning in urban scenarios.

As mentioned before, supervised learning can significantly reduce training time for a control system. For this reason, Xia et al. [105] presented a vehicle control algorithm based on Q-learning combined with a pre-training phase based on expert demonstration. A filtered experience replay, where previous experiences were stored while poor performances were eliminated, was used to improve the convergence during training. The use of pre-training and filtered experience replay was shown to not only improve final performance of the control policy, but also speed up convergence by up to 71%. Comparing two learning algorithms for lane keeping, Sallab et al. [106] investigated the effect of continuous and discretized action spaces. A DQN was chosen for discrete action spaces, while an actor-critic algorithm was used for continuous action values. The two networks were trained and evaluated driving around a simulated race track, where their goal was to complete the track while staying near the center of the lane. The ability to utilize continuous action values resulted in significantly stronger performance with the actor-critic algorithm, enabling a much smoother control policy. In contrast, the DQN algorithm struggled to stay near the center of the lane, especially on curved roads. These results demonstrated the advantages of using continuous action values.

Using vision to control the autonomous vehicle, Zhang et al. [107] presented their supervised learning algorithm, SafeDAgger, based on the Dataset Aggregation (DAgger) [108] imitation learning technique. In DAgger, the first phase of training consists of traditional supervised learning where the model learns from a training set collected from an expert completing a task. Next, the trained policy performs at the given task and these experiences are collected into a new data set. The expert is then asked to label the newly collected data with the correct actions. The labeled data sets are then added to the original data set to create an augmented data set, which is used to further train the model. The process is then repeated in the DAgger training framework. This approach is used to overcome one of the limitations of supervised learning for control policies. When using a supervised learning-based algorithm, as the trained model makes slight errors, it veers away from its known operational environment and such errors can compound over time. For instance, when steering a vehicle, the model may have learned from a data set where the expert is always in (or near) the center of the lane. However, when the trained network is steering the vehicle, small errors in the predicted steering angle can slowly bring the vehicle away from the center of the lane. If the network has never seen training examples where the vehicle is veering away from the center of the lane, it will not know how to correct its own mistakes. Instead, in the DAgger framework, the network can make these mistakes, and the examples are then labeled with correct actions by the expert, and the next training phase then allows the network to learn to correct its mistakes. This results in a control policy more robust to small errors, however, it introduces further complexity and cost in accessing the expert for further ground truth queries. The framework was later extended in SafeDAgger, by developing a method that estimates, for the current state, whether the trained policy is likely to deviate from the reference (expert) policy. If it is likely to deviate more than a specified threshold, the reference policy is used to control the vehicle instead. The authors used the SafeDAgger method to train a CNN to control the lateral and longitudinal actions of an autonomous vehicle while driving on a test track. Three algorithms were compared: traditional supervised learning, DAgger, and SafeDAgger, and SafeDAgger was found to perform the best. Another work exploring a DAgger-type framework for autonomous driving was presented by Pan et al. [109]. The method used a model predictive control algorithm to provide the reference policy using high cost sensors, while the learned policy was modeled by a CNN with low cost camera sensors only. The advantage of this approach was that the reference policy can use highly accurate sensing data, while the trained policy can be deployed using low cost sensors only. The approach was evaluated experimentally on a sub-scale vehicle driving at high speeds around a dirt track, where the trained policy demonstrated it could operate the vehicle using the low cost sensors only.

In a move away from direct perception methods, Wang et al. [110] presented an object-centric deep learning control policy for autonomous vehicles. The method used a CNN which identified salient features in the image, and these features were then used to output a control action, which was discretized to {left, straight, right, fast, slow, stop} and then executed by a PID controller. Simulation experiments in Grand Theft Auto V demonstrated the proposed approach outperforms control policies without any attention or those based on heuristic feature selection. In another work, Porav and Newman [111] presented a collision avoidance system using deep reinforcement learning. The proposed method used a variational autoencoder coupled with a RNN to predict continuous actions for steering and braking using semantically segmented images as input. Compared to the collision avoidance system with braking only by Chae et al. [98], a significant improvement in reducing collision rates for TTC values below 1.5 s was demonstrated, with up to 60% reduction in collision rates.

These works demonstrate that a DNN can be trained for simple driving tasks such as staying in its lane, following a vehicle, or collision avoidance. However, driving not only consists of multiple such tasks, but also requires outside context of a higher goal as well. Humans drive vehicles with the aim to start from point A and arrive at point B. Therefore, having the ability to only follow a single lane is not enough to fully automate this process. For example, it has been reported that end-to-end road following algorithms will oscillate between two different driving directions when coming to a fork in the road [83]. Therefore, the networks should also be able to learn to utilize a higher level goal to take correct turns and arrive at the target destination. Aiming to provide such context awareness for autonomous vehicles, Hecker et al. [112] used a route plan as an additional input to a CNN-based control policy. The network was trained with supervised learning, where the human driver was following a route plan. These driving actions, along with the route plan, were then included in the training set. Although no live testing with the network in control of the car was completed, qualitative testing with example images from the collected data set suggested the model was learning to follow a given route. In a similar approach, Codevilla et al. [113] used navigation commands as an additional input to the network. This allows the human or a route planner to tell the autonomous vehicle where it should turn at intersections, and the trained control policy simply follows the given route. A network architecture which shared the initial layers, but had different sub-modules (feedforward layers) for each navigational command, was found to work best for this approach. When given a high-level navigational command (e.g., turn left), the algorithm uses the final feedforward layers specifically trained for that command. The trained model was evaluated both in simulation and in the real world using a sub-scale vehicle. The proposed approach was shown to successfully learn to follow high-level navigational commands.

In contrast to the widely popular end-to-end techniques, Waymo recently presented their autonomous driving system, ChauffeurNet [114]. ChauffeurNet uses what the authors refer to as mid-to-mid learning, where the input to the network is a pre-processed top-down view of the surrounding area and salient objects, while the output is a high-level command describing target way point, vehicle heading, and velocity. The final control actions are then provided by a low-level controller. The use of mid-level representation allows easy mixing of simulated and real input data, making transfer from simulation to the real world easier. Furthermore, this means that the model can learn general driving behavior, without the burden of learning perception and low-level control tasks. In order to avoid dangerous behavior, the training set was further augmented by synthesizing examples of the vehicle in incorrect lane positions or veering off the road, which enabled the network to learn how to correct the vehicle from mistakes. The trained model was evaluated in simulation and real-world experiments, demonstrating desirable driving behavior in both cases. The authors noted that the use of synthesized data, augmented losses, and a mid-to-mid learning approach were key to allowing the model to learn to drive the vehicle while avoiding undesirable behaviors.

A summary of the full vehicle control techniques presented in this chapter can be seen in Table 3.3. In contrast to the previous sections, a variety of learning and implementation approaches were used in this section. Machine learning approaches have demonstrated strong results for driving tasks such as lane keeping or vehicle following. While these results are important, for fully automated driving, more complex tasks need to be considered. The network should be able to consider low-level goals (e.g., lane keeping, vehicle following) alongside high-level goals (e.g., route following, traffic lights). Early research has been carried out integrating high-level context to the machine learning models, but these models are still far from human-level performance.

3.2   RESEARCH CHALLENGES

The discussion in previous sections has shown that important results have been achieved in the field of autonomous vehicle control using deep learning techniques. However, research in this area is on-going, and there exists multiple research challenges that prevent these systems from achieving the performance levels required for commercial deployment. In this section, we will discuss some key research challenges that need to be overcome before the systems are ready to be deployed. It is worth remembering that besides these research challenges, there are also more general challenges preventing the deployment of autonomous vehicles such as legislative, user acceptance, and economic challenges. However, since the focus here is on the research challenges, for information on more general challenges we refer the interested readers to the discussions in [115–120].

3.2.1   COMPUTATION

Computation requirements are an issue for any deep learning solution due to the vast amount of training data (supervised learning) or training experience (reinforcement learning) required. Although the increase in parallelisable computational power has (partly) sparked the recent interest in deep learning, with deeper models and more complex problems the computational requirements are becoming an obstacle for deployment of deep neural network driven autonomous vehicles [121–123]. Furthermore, the sample inefficiency, which many state-of-the-art deep reinforcement learning algorithms suffer from [124], means that a vast number of training episodes are required to converge to an optimal control policy. Since training the policy in the real world is often unfeasible, simulation is used as a solution. However, this results in high computation requirements for reinforcement learning.

Table 3.3: A comparison of full vehicle control techniques

Ref.	Learning Strategy	Network	Inputs	Outputs	Pros	Cons
[104]	Supervised learning	RNN	Relative distance, relative velocity	Desired acceleration	Adapts to driving styles of other vehicles	Only considers a simplified roundabout scenario
[105]	Supervised reinforcement learning	Feedforward network with two hidden layers	Not mentioned	Steering, acceleration, braking	Fast training	Unstable (can steer off the road)
[106]	Reinforcement learning	Feedforward/ Actor-Critic Network	Position in lane, velocity	Steering, gear, brake, and acceleration values (discretized for DQN)	Continuous policy provides smooth steering	Simple simulation environment
[107]	Supervised learning	CNN/Feedforward	Simulated camera image	Steering angle, binary braking decision	Estimates safety of the policy in any given state, DAgger provides robustness to compounding errors	Simple simulation environment, simplified longitudinal output
[109]	Supervised learning	CNN	Camera image	Steering and throttle	High-speed driving, learns to drive on low cost cameras, robustness of DAgger to compounding errors	Trained only for elliptical race tracks with no other vehicles, requires iteratively building the dataset with the reference policy
[110]	Supervised learning	CNN	Image	nine discrete actions for motion	Object-centric policy provides attention to important objects	Highly simplified action space
[111]	Reinforcement learning	VAE-RNN	Semantically segmented image	Steering, acceleration	Improves collision rates over braking only policies	Only considers imminent collision scenarios
[113]	Supervised learning	CNN	Camera image, navigational command	Steering angle, acceleration	Takes navigational commands into account, generalizes to new environments	Occasionally fails to take correct turn on first attempt
[112]	Supervised learning	CNN	360° view camera image, route plan	Steering angle, velocity	Takes route plan into account	Lack of live testing
[114]	Supervised learning	CNN-RNN	Pre-processed top-down image of surroundings	Heading, velocity, waypoint	Ease of transfer from simulation to the real world, robust to deviations from trajectory	Can output waypoints which make turns infeasible, can be over aggressive with other vehicles in new scenarios

3.2.2 NETWORK ARCHITECTURES

The current model design in deep learning research often depends largely on rules of thumb and heuristic techniques. Currently, there are no efficient techniques for choosing the optimal network hyperparameters, and trial-and-error is often used instead [125, 126]. A common method for hyperparameter tuning is Grid Search [127, 128], where a range of values for each hyperparameter are chosen, and training is completed for different hyperparameter setups. The best performing hyperparameter values are then chosen for the final design. However, this heuristic approach to network parameter tuning is very time intensive and error prone [129, 130]. This hinders the research process, as a significant part of the research project can be used for tuning the model parameters to increase the accuracy of the model. This also ties into the first research challenge, as the need to re-train the model with slightly changed parameters increases the amount of computation needed to arrive at the final model design. There is ongoing research in automated neural architecture search methods looking to solve this problem [131–135].

3.2.3 GOAL SPECIFICATION

Goal specification is an issue specifically in reinforcement learning. The goal of any reinforcement learning model is to learn a control policy which maximizes the rewards. Therefore, the reward function is the means by which the designer describes the desired behavior the model should learn [136, 137]. However, this can be difficult to do effectively for complex tasks. In automated driving, multiple conflicting objects need to be balanced against one another. A poorly designed reward function cannot only hinder the convergence of the model, but also cause unexpected model behavior. Therefore, care needs to be taken that the reward function adequately represents the desired behavior, and no unexpected ways to exploit the reward function exist. A classic example of such reward hacking is a robot trained for ball paddling, where the reward function is the distance between the ball and a desired height [138]. Although the intended behavior for the robot is to hit the ball such that it reaches the desired height consistently, the robot may learn to exploit the reward function by resting the ball on the paddle and raising the paddle and ball together to the target height. Therefore, the reward function should be designed such that there are no unintended ways of obtaining high rewards [139].

3.2.4 GENERALIZATION

Generalization is a key challenge when working in any complex operational environment. This is especially true in autonomous driving, as the operational environment is very complex, and can differ drastically between different scenarios. Differences in road type (urban vs. highway), weather (dry vs. snow), regional road laws, local signs and road marks, traffic, and host vehicle parameters can all have a significant effect on the driving behavior required, and the autonomous vehicle must be able to adapt to these different environments. However, due to the complexity of the operational environment, testing the system in all possible scenarios is unfeasible [140, 141]. Therefore, the model needs to be able to generalize to previously unseen scenarios. There are tools for training DNNs, which aim to avoid overfitting to the training data. Early stopping uses a validation data set, a separate data set from the training set not used for weight updates, which is used to observe the performance of the model on unseen data. If the validation performance begins to degrade while the training set score increases, the training should be stopped as the model is beginning to overfit. Regularization techniques are another set of tools to reduce test errors, although sometimes at the cost of an increased training error. Regularization by constraining the weights of the network or dropping some neurons out during training ensures trends over the whole training set are learned and complex co-adaptations between neurons are prevented [142–146].

3.2.5   VERIFICATION AND VALIDATION

Verifying the model performance is important for any real world vehicle application. Currently, the trend in the field is to evaluate the model performance in simulation, due to the high cost and time requirements of field testing [147]. While simulation allows for faster testing of novel ideas and easier testing of safety-critical scenarios, it must be kept in mind that the use of simulation will introduce various model inaccuracies. Accurately modeling physical phenomenon such as friction and contact can be difficult. However, these inaccuracies mean that a model trained in simulation may not work in the real world. This is exacerbated by the use of vision, as creating an adequately realistic graphical simulator can be difficult. This leads to the model overfitting to the simulation environment, and failing to generalize to the differences in the real-world environment. The transfer from simulation to real-world training is currently an active research area, with many promising approaches for solving this problem such as domain rrandomization [148–150], domain adaption [151–154], and iterative learning control [155–157]. Furthermore, approaches such as the mid-to-mid learning in ChauffeurNet [114] can be used as an alternative work around to this issue. Regardless of the training method, extensive real-world testing of any autonomous vehicle application will be required to ensure the model performs adequately in its intended operational environment [158].

3.2.6   SAFETY AND INTERPRETABILITY

The safety of the deep learning models is perhaps the most critical issue hindering the deployment of learned autonomous vehicles. A serious mistake or malfunction in an autonomous vehicle can cause death or serious harm to people. Therefore, in a safety-critical system such as an autonomous vehicle, the safety of each sub-component as well as the overall system must be validated to a high assurance level. However, this is challenging where deep neural networks are used, as the systems are opaque in nature. This is referred to as the black box problem [159]. While we can test the trained models in targeted testing to observe model performance, the complexity of the deep learning models means that understanding what the system has learned and what rules it follows is practically impossible. This means that traditional safety validation techniques are less useful where deep neural networks are used. Therefore, new techniques, or modifications to existing techniques, must be found to address this drawback of deep learning models [139, 160]. For these reasons, we will explore issues in safety validation of deep neural networks and possible solutions in the next chapter.

3.3 SUMMARY

A review of state-of-the-art techniques in autonomous vehicle control via deep learning was given in this chapter. The review was broken into three sections: lateral, longitudinal, and full vehicle control. Currently, the trend in lateral control algorithms has been to use supervised learning from images, while longitudinal control algorithms tend to favor using reinforcement learning from lower dimensional states. On the other hand, in full vehicle control, a wide variety of techniques have been proposed. Impressive results have been achieved in this field, and research is beginning to move away from learning simple driving tasks, to learning to drive using high-level context while also considering low-level objectives. The main research issues for the future prospects of the field were discussed, and potential solutions were introduced. The most critical research challenge, safety and interpretability of DNNs, was identified and will be explored further in the next chapter.

CHAPTER 4

Safety Validation of Neural Networks

The safety-critical nature of autonomous vehicles means that safety is a critical objective in the design of the system and must be ensured to a high assurance level. In typical automotive applications this is done through functional safety validation, where the system is proven to be safe to a certifiable level, following the guidelines of a standard such as ISO 26262 [161]. The lack of adequate functional safety validation techniques which can ensure safety to a certifiable level is currently a significant obstacle for the implementation of deep learning systems in autonomous vehicles. Deep neural networks introduce a number of challenges which hinder the application of traditional verification and validation techniques. First, the opaque nature of deep neural networks limits their interpretability. With the field moving to more complex and deeper models, interpreting what the model has learned and how it makes its decisions is no longer feasible [159]. Second, online learning systems have the capability to self-optimize their performance by changing their internal structure during operation. This leads to further difficulty for functional safety validation when online learning is utilized, as these systems can no longer be fully validated offline [162–166]. Therefore, methods for run-time validation must also be considered for adaptive systems. Third, the complex nature of the operational environment for autonomous vehicles means that it is impossible to test the system for all possible eventualities [140, 141]. Therefore, the validation process must provide proof that the system has learned rules which ensure safe behavior is maintained even in previously unseen scenarios. Finally, deep neural networks introduce new vulnerabilities, such as malicious inputs through adversarial examples. Adversarial examples add small perturbations to images, which cause DNNs to misclassify them with high confidence [167–171]. These factors make the functional safety validation of deep learning systems significantly more difficult when compared to traditional software systems. New functional safety validation approaches must be found, or existing methods adapted, to move the field toward an adequate safety validation process for deep learning systems. For these reasons, this chapter is dedicated to reviewing a variety of safety validation approaches and analyzing their applicability to deep learning-based systems for autonomous vehicles.

4.1   VALIDATION TECHNIQUES

4.1.1   FORMAL METHODS

Formal methods are a class of techniques which rely on formal logic and discrete mathematics for mathematical specification, design, and verification of systems [172–175]. Applying formal methods for verification and validation of software can be a rigorous and complex task, but can provide high safety assurance [176]. Typically, a mathematical model of the system is used to apply formal methods to prove that the system is complete and correct [177]. However, due to the opacity of neural network systems, the applicability of formal methods to neural networks is currently limited. Nevertheless, existing formal methods can provide limited insight when applied to specific types of neural networks.

Rule extraction is a technique used to extract high-level rules regarding the operation of the network, where the rules are examined to determine the acceptability and safety of the network. During development, these extracted rules could be used to provide further understanding of what the network is learning. In some types of networks, rule insertion can also be used to insert rules into the network to ensure the desired behavior. Such methods have, for example, been shown to work in hybrid neuro-fuzzy systems [178, 179]. Methods extracting guarantees from neural networks have been investigated as well. For instance, Pulina and Tacchella [180] encoded a neural network as a set of constraints which could then be solved by a Boolean Satisfiability Problem solver. This abstraction allowed formal methods to be applied to the network and provided some guarantee of the networks outputs in the desired neighborhoods. However, the technique was only tested on a six-neuron network and the sigmoid activation function was approximated using constraints, limiting the potential insight into its efficacy on practical deep neural networks. This method was further extended by Katz et al. [181] in the Reluplex algorithm. The method is based on the Simplex verification method, extended to work on ReLU activation functions by utilizing a search tree instead of the search path approach used in Simplex. While this system specializes in networks with ReLU neurons, it was shown to provide some behavioral guarantees in networks with up to 300 neurons. On the other hand, Huang et al. [182] proposed a framework for verifying the robustness of feedforward multi-layer neural networks to adversarial examples based on Satisfiability Modulo Theory. The proposed method searches for adversarial examples for the network by analyzing the network layer by layer. If no adversarial examples are found, the network can be verified as safe. However, if such adversarial examples are found, they can be used to train the network thus improving the robustness of the network. The main weakness of the method was that the verification process increases exponentially in computational complexity with the increasing number of features, limiting the scalability of the method. Moreover, the technique relies on various assumptions (e.g., assuming that the output for a given input is mainly dependent on a small subset of neurons). Overall, the trend seen in current formal verification methods is that some guarantees can be obtained for smaller networks, but the techniques are limited by their reliance on assumptions and lack of scalability to larger networks.

There are a number of identifiable challenges which limit the current use of formal methods and they might become more powerful tools for neural network validation if there are improvements in environment modeling, formal specification, and interpretability of the networks. Environment modeling poses a challenge, since in comparison to traditional formal verification methods (where the environment can be well defined and often even over-approximated), in neural network systems the environment can be too complex to precisely define. Therefore, the uncertainty of the model must be taken into account in the verification and validation process. One approach is to identify assumptions made about the environment at design time, and then monitor the validity of these assumptions at run-time. A further challenge is to provide accurate formal specifications for the system. This is challenging in domains where the operational environment of the neural network system is very complex. The formal specification must accurately describe the desired and undesired behavior of the system. This can be made easier by focusing on creating a system level description to specify the end-to-end behavior of the entire neural network, rather than focusing on creating a component level specification for each element of the system. For instance, for an autonomous vehicle, the behavior should not be specified at the level of the AI system itself, but overall vehicle behavior such as minimum distance from other obstacles, which can be measured and defined more easily [175].

4.1.2   RUN-TIME MONITORING

Run-time monitoring can be used to monitor the real-time behavior of the system to ensure safe operation. It is well suited to complex cyber-physical systems where exhaustive offline testing is often impossible due to adaptive, non-deterministic and/or the near infinite state nature of these systems [163]. Run-time monitoring is used to ensure that the system operates in a pre-defined safe boundary. Additionally, run-time monitoring can be used to ensure system stability and identify non-convergent behavior using methods based on the Lyapunov stability theory [183]. The core idea of run-time monitoring is to identify potential safety problems arising in the system before they can cause accidents. Therefore, in addition to monitoring techniques, thought must also be given to recovery techniques in case the safety monitors are violated. The first approach is to shut down the system and stop operation. However, this does not allow the system to recover safe operation, and moreover shutting down an autonomous agent such as a vehicle can be dangerous in some environments. A second approach is utilizing a second recovery controller. This recovery controller would be a controller which can ensure safe operation, even if it is with degraded performance and reduced capabilities. In autonomous vehicle systems, this could be a traditional rule-based controller which has lower performance but for which safety can be more easily guaranteed. This type of approach is also referred to as run-time monitoring and switching. A third approach utilizes software safety cages which limit the outputs of the system within safe bounds.

It should be noted that run-time monitoring methods are very domain specific, and their applicability depends on the availability of effective safety metrics and recovery techniques.

Moreover, run-time monitoring and validation should also account for uncertainties in state observations. These can be caused, for example, by sensor inaccuracies and defects or software faults. Therefore, run-time monitoring techniques must accurately account for the impact of hardware and software faults, otherwise the safety violations may not be identified in time to prevent accidents. One approach for considering these uncertainties is the use of probabilistic models. In this approach, the behavior of the system and its environment is described by a likelihood based on the confidence in the safety monitors' observations. The safety monitors use these probabilistic models to provide a likelihood that the system is behaving according to its specification [163]. There have been a number of examples of such probabilistic run-time monitors proposed in the literature; for instance, see [184–189]. A further drawback of run-time monitoring methods is the increased computational overhead in systems where checking complicated properties is involved. In some cases it may be acceptable to lower the accuracy of the safety monitors to reduce the increased computational overhead for example, by using a control-theoretic approach to implement a trade-off between accuracy and computational overhead [190].

4.1.3   SOFTWARE SAFETY CAGES

Software safety cages can be used to improve the safety of a system by limiting the system outputs to a safe envelope. In their simplest form, software safety cages are simply hard upper and/or lower limits on the system output. However, when combined with run-time monitoring methods, the software safety cages can be dynamic. By using observations from the safety monitors as context, the software safety cages can limit the output of the neural network system dynamically based on the current state. For example, in an autonomous vehicle, this approach can be used to prevent acceleration or force braking of the vehicle if the vehicle is within a critical distance from another vehicle, regardless of the outputs of the neural network controller. In this approach, potentially dangerous outputs of the neural network can be prevented by limiting their outputs during critical scenarios. Therefore, the safety validation requirements on the neural network can be relaxed, given that the software safety cages can be validated with high assurance.

Heckemann et al. [191] proposed a framework which utilizes context aware safety cages to ensure the safety of complex and adaptive software algorithms in automotive applications. The safety cages in this framework evaluate the hazards and the driving situation and assign an Automotive Safety Integrity Level (ASIL) to their combination based on the estimated levels of exposure, severity, and controllability. For instance, an emergency braking maneuver has a higher severity in a high-speed driving situation. Therefore, the software safety cages can use their context awareness to check the plausibility of function outputs and restrict the system outputs depending on the current vehicle state. For example, although an emergency braking maneuver might be allowed in an urban scenario, it might be restricted to limited braking power when driving at high speeds on a highway. Furthermore, since the large number of inputs in such a system would cause the safety cage to be very complex, the overall vehicle safety cage is split into multiple smaller safety cages. Each individual safety cage addresses a certain aspect of the system. Actuator safety cages are focused on safety-related actuators which influence the vehicle's lateral or longitudinal dynamics. Functional safety cages are focused on specific functions of the vehicle such as automatic parking or overtaking. By defining specific domains for each safety cage, the safety cages can be tailored to that domain. A further benefit of this architecture is that it introduces functional safety cages with low complexity that oversee each function of the vehicle. Since these safety cages are more transparent and less complex than the neural network controllers they would oversee, they can be validated for safety using traditional safety validation methods such as formal methods supplemented with testing data. Therefore, a high ASIL classification can be given to these safety cages while a low ASIL classification is given to the neural network controller, thereby shifting the validation burden from the controller to the safety cages [192].

4.1.4   CROSS VALIDATION

Cross validation is an approach to provide reliability through redundancy [174]. The core idea is to use an ensemble of diverse neural networks which each independently estimate an output from the same inputs [193]. The outputs of each network can then be checked against one another to validate the system output. The validation is therefore designed into the system architecture, rather than as a separate process. However, the main purpose of cross validation is to increase confidence in the output, but it does not give full assurance of the system validity since systematic errors could occur in each of the networks if they are not sufficiently diverse. The use of ensembles can provide better performance if the diversity of the networks can mitigate the effects of errors. Since any network will have errors in at least some subset of its input space, diverse networks in the ensemble offer complementary responses, such that (weighted) averaging of the input can be used to get a better prediction of the desired output [194]. Alternatively, instead of all of the networks affecting the final output, the variance in the ensemble outputs could be used to estimate the uncertainty in the output of the neural network controller.

In order for cross validation to be effective, there must be some diversity in the neural networks in the ensemble [195, 196]. Clearly, if each network has the same output for all possible inputs, no additional information is gained [197]. Therefore, some method of creating diversity in the ensemble networks must be found. One approach is to use multiple networks with different topologies for the same task. However, it may be challenging to define multiple diverse network architectures that can provide adequate performance for the given task. Another approach is to use multiple networks with the same topology, but each trained with different training data. This has the advantage that one optimal network architecture can be used for the whole ensemble. However, since each network requires diverse training data, the quantity of data required is significantly increased. Regardless of the approach for creating diversity, the performance of each network must be carefully evaluated. While we do want to promote diversity in the outputs of each network, this should not come at the price of significantly reduced performance. Therefore, the performance of each neural network should be evaluated on the test data, and inadequately performing networks should be excluded from the final ensemble.

Investigating the use of cross validation for testing the robustness of deep neural networks, Pei et al. [198] proposed an automated testing system for deep neural networks called DeepXplore. Pei et al. introduced neuron coverage as a measure for the extent to which the network's logic space has been tested. Neuron coverage calculates the fraction of neurons in the neural network which were activated during the testing process. A neuron is considered to have been activated if its output exceeds a pre-defined threshold limit. Using this measure of neuron coverage, DeepXplore can systematically test the deep neural network for erroneous behavior and synthesize new inputs with the aim to maximize neuron coverage, thereby ensuring that all parts of the network are covered during the testing process. The system compares inputs between multiple deep neural networks built for the same task and cross validates the outputs to identify erroneous behavior. For example, given an ensemble of three networks built to steer a vehicle, if two of the networks decide to turn right while one decides to turn left, the latter one is assumed to be behaving erroneously. This has the advantage that no manual labeling for synthesized test inputs is required. However, the disadvantage is that erroneous behavior can only be detected if at least one network outputs different decisions than the other networks in the ensemble. If all networks show the same erroneous behavior, DeepXplore will fail to identify it. Also, the system assumes that the decision made by the majority of the networks is correct, which may not always be the case. To test the deep neural networks for erroneous behavior, DeepXplore generates synthesized test inputs which attempt to maximize the neuron coverage in the networks and produce differential behaviors between them. Therefore, the system attempts to jointly optimize these two objectives when generating test inputs. The system was tested using five different widely used public datasets: MNIST [199], ImageNet [200], Udacity Challenge [201], Contagio/VirusTotal [202, 203], and Drebin [204, 205]. DeepXplore was evaluated on 3 deep neural networks for each dataset, for a total of 15 networks. In the image recognition datasets (MNIST, ImageNet, and Udacity), the synthesized test inputs were created by applying transformations to the original images in the dataset. Three transformations were leveraged: (1) changing lighting conditions, (2) occluding part of the image with a rectangle to simulate an attacker blocking a part of the image, and (3) occluding the image with multiple small rectangles to simulate the effect of dirt on the camera lens. DeepXplore was shown to find thousands of examples of erroneous behavior, such as an autonomous vehicle attempting to crash into a guard rail in the Udacity dataset. Moreover, it was shown that the error inducing test cases could be used as training data to improve the robustness of the networks. This was shown to achieve 1–3% improved accuracy over adversarial and random training data augmentation methods.

4.1.5   VISUALIZATION

Visualization can be a useful tool to improve the interpretability of neural networks for verification and validation practitioners. Visualization techniques can be used to transform data to forms that humans can interpret more easily. There are already visualization tools and techniques in traditional software verification and validation which are used to create visual representations of data to improve interpretability [174]. Visualization techniques for neural networks could, for example, be used to create graphical representations of changes in weights or internal connections in the network, plots of error functions over the learning process to improve understanding of the decision making process and the learning process during training. Therefore, these representations can give greater insight into the structure of the neural network, including weights and biases and their respective changes during training [206, 207].

Visualization tools have gained significant interest recently for the interpretation of CNN learning. One popular technique is Activation Maximization which provides insight into which features the CNN classifiers have learned to relate to different classifications. Activation Maximization synthesizes an image which maximizes the output for a given neuron or output. Another use for this technique is to create an adversarial example, which is an image that is unrecognizable to humans but outputs a high confidence classification by the CNN [167]. However, Simonyan et al. [208] introduced a regularization technique into this process, which created more recognizable images, giving insight into the kind of features the CNN classifier was looking for in the specific classes. Yosinski et al. [209] further extended this method with better regularization techniques as well as investigation of neurons at all layers, rather than limiting the study to the output neurons. This showed that neurons at different layers were learning different features, with higher layers learning more complex and abstract features (e.g., faces, wheels, eyes) while the lower layers were learning more basic features (e.g., edges and corners). Therefore, this type of visualization shows great potential to improve interpretability of CNNs, as they can provide significant insight into what the neural network has learned [209]. For instance, such visualization techniques were used by Bojarski et al. [87, 210] in the NVIDIA PilotNet project to visualize the internal state of the CNN used for steering. By studying the activations within different layers of the trained CNN, they were able to gain a better understanding of what features the neural network had learned to recognize. The analysis showed that even with only the human steering angles as training input, the CNN had learned to recognize useful road features such as the edges of the road. The authors also investigated the activations in the network when given an image with no road as input and found that the activations of the two feature maps mostly contained noise, indicating that the CNN found no useful features in the image. Therefore, the CNN only learned to recognize features that were useful for its task, such as road-related features.

Another useful visualization technique investigates the importance of different neurons in creating predictions. By analyzing the gradients flowing into the last convolutional layer in the CNN, the contribution of each neuron to the final prediction can be determined. This information can then be used to create heat maps which visualize the parts of the input image that contribute most to the neural network's output. Similar heat maps can also be produced by occluding a region of the image and observing the change in the classification output. By repeating this process iteratively over all regions of the input image, the contribution of each image region to the output can be plotted [211].

4.1.6   BLACK BOX TESTING

Black box testing refers to test methods where the internal structure, design, and implementation of the software is unknown to the tester. The software is therefore only being tested for its outputs to specified inputs and the general usability of the system. The aim of black box testing is to investigate interface errors, initialization and termination errors, as well as performance. This has the advantage that the testers do not need to be familiar with the internal structure of the software, therefore the test can be carried out by an independent body from the developers, removing developer-bias from testing. Due to the inherent opaqueness of neural networks, black box testing is likely to play a critical role in the safety validation of these systems [141].

For neural network controllers in automotive applications, these black box testing techniques will include targeted testing of the learned function. This will include testing the vehicle controller in a number of critical scenarios (e.g., emergency braking at high speeds, avoiding pedestrians on the road, etc.) with a variety of parameters (e.g., distance to lead vehicle at the beginning of emergency braking). Using black box testing with a wide variety of test scenarios will allow for a better understanding of its performance and generalization capabilities. However, it is unlikely to be viable to carry out all of this testing through field testing alone. Therefore, simulation as well as manipulated real data will play a critical role in the black box testing. A detailed analysis of the impact of synthetic data sources should be carried out to understand the reliability of the results. Therefore, the black box testing for learned autonomous vehicles will likely include evidence from field tests, simulation, and statistical extrapolation. Using a combination of these sources, the aim should be to obtain the best possible coverage of the desired operational environment, in order to ensure the results are representative of the system's actual level of performance [141].

Grey box testing is also a promising validation framework for deep learning systems. Grey box testing combines partial knowledge of the internal mechanisms of a software system with black box testing methods [212]. The benefit of grey box testing is that the testing still involves investigating the inputs and outputs using the straightforward and unbiased nature of black box testing methods, however introducing partial knowledge of the system's internal mechanisms allows the designer to develop test methods better suited to the specific system. For instance, a grey box testing framework for deep neural network-driven autonomous vehicles was investigated by Tian et al. [213]. By combining black box testing methods and the measure of neuron coverage suggested by Pei et al. [198], Tian et al. introduced DeepTest, an automated testing tool for deep neural network-driven vehicles. DeepTest generates synthesized inputs using nine different image transformations: changing brightness, changing contrast, translation, scaling, horizontal shearing, rotation, blurring, fog effect, and rain effect. However, instead of using majority voting in an ensemble of deep neural networks like DeepXplore, DeepTest compares the output in the transformed image to the labeled correct output in the original image to identify erroneous behavior. The disadvantage of this method is that false positives can occur in situations where the new output is in fact. For example, in an image which has been transformed by horizontal shearing, a different steering angle may be required to stay on the road compared to the original image. The DeepTest framework was applied to three top scoring deep neural networks from the Udacity self-driving challenge, where deep learning was used to predict steering angles given an input image from a front-facing camera mounted on a vehicle. The results showed that the system can find thousands of erroneous behavior cases for each network. However, due to the above-mentioned drawback, a trade-off between the number of identified erroneous cases and frequency of false positives needs to be made. For instance, in one test case among 6339 reported erroneous cases, 130 false positives were identified through manual investigation. The drawback is that false positives are difficult to identify and currently require manual checking. Furthermore, similar to DeepXplore, identified erroneous test cases can be used to train the system for improved robustness.

4.2   DISCUSSION

Deep neural networks have a number of challenges which render traditional verification and validation methods ineffective. The verification and validation of these systems is challenging due to the high complexity and nonlinearity inherent to them [214]. The lack of interpretability in these systems means that the systems cannot be modeled mathematically, limiting the effectiveness of formal methods. Also, due to the large parameter space in complex tasks such as driving, testing the system for all possible inputs is impossible which limits the effectiveness of black box testing methods [215]. The validation of online learning systems has further challenges due to the potential ability to continue learning and change their internal structure over time after deployment. It is therefore difficult to ensure that the desired behavior, even if proven at design time, does not drift to undesired behavior due to the adaptation of the system to the operation environment. These adaptations, while typically beneficial to system performance, could lead to undesirable behavior due to poor design or poor inputs (e.g., erroneous inputs from faulty sensors). Therefore, static verification and validation methods are not enough for online learners. This means that verification and validation for online learners require run-time methods [160]. Hence, run-time monitoring and software safety cage methods will play a critical role in ensuring the safety of online learning systems. Furthermore, validating the software algorithm itself is not enough hence emphasis must also be given to validating the training data. Varshney and Alemzadeh [216] noted that the influence of training data on a neural network system's behavior can be as impactful as the algorithm itself. Therefore, efforts must be made to ensure that the training data is adequate for training the system for its desired task. The training data validation must include consideration for the volume of data, coverage of critical situations, minimization of unknown critical situations, and representativeness of the operational environment [141].

Another challenge for autonomous vehicle applications is the adequacy of current standards such as ISO 26262 for safety validation of neural network systems. These safety standards have helped develop industry practices to address safety in a systematic way. However, Salay et al. [217] noted that ISO 26262 in its current form does not address machine learning methods adequately. Salay et al. identified five factors from machine learning which will likely impact ISO 26262 and require changes in the standard: (i) identifying hazards, (ii) faults and failure modes, (iii) use of training sets, (iv) level of machine learning usage, and (v) required software techniques. First, identifying hazards, as specified by ISO 26262 currently, is an issue as machine learning can create new types of hazards which might not necessarily fit the definition of hazards as given by the standard. Therefore, the definition of hazards in ISO 26262 should be revised to also consider harm potentially caused by complex behavioral interactions between the autonomous vehicle and humans that are not due to a system malfunction. Second, faults and failure modes will be further affected by machine learning methods as they will introduce machine learning-specific faults in network topology, learning algorithm, or training set, which will need to be addressed by ISO 26262. Therefore, ISO 26262 should require the use of fault detection tools and techniques which take into account the unique features of machine learning. Third, the use of training sets is problematic from the perspective of ISO 26262 certification, as it breaks an assumption made by the standard that component behavior is fully specified and each refinement can be verified with respect to its specification. However, where training sets are used in place of specifications, this assumption is not valid (as training sets are inherently incomplete). Therefore, training set coverage should be considered instead of completeness. System specification may be an issue for systems with more advanced functionality, such as perception of the environment, as these may be inherently unspecifiable. Hence, complete specification requirement in ISO 26262 should be relaxed. Fourth, the level of machine learning usage could be a further issue as ISO 26262 assumes the software can be defined as an architecture consisting of components and their interactions in a hierarchical structure. However, this is not always the case for machine learning systems. For example, in end-to-end systems there are no sub-components or hierarchical structure and therefore these systems challenge the assumptions in the standard. Moreover, ISO 26262 mandates use of modularity principles such as restricting the size of components and maximizing the cohesion within a component, which could be problematic for machine learning components that lack transparency and therefore cannot apply these principles. Finally, the required software techniques in ISO 26262 are a further challenge for machine learning methods as many of them assume that an imperative programming language is being used. Salay et al. assessed the 75 software techniques required by ISO 26262 and found that approximately 40% of these are not applicable to machine learning, while the rest are either directly applicable or can be applied if adapted in some way. Therefore, ISO 26262 software techniques should perhaps focus more on the intent than on specific details. Additionally, more work is required on the software development requirements for machine learning systems to provide requirement criteria for issues such as training, validation, test datasets, training data pre-processing, and management of large data sets [218].

It is clear that verification and validation efforts should be carried out throughout the lifecycle of the system. The process should start with a clear validation plan with the requirements of the network in mind. To help build a case for safety of the system, approaches such as Goal Structuring Notation (GSN) [219] could be utilized. GSN specifies a set of goals for the system which, if fulfilled, can be used as evidence for arguing the safety case of the system. See Fig. 4.1 for an example of a general GSN format for a neural network system suggested by Kurd et al. [179]. It should be noted that for a complete GSN, additional hazards specific to the intended application domain should be considered and used to build additional safety goals. During the design phase, safety of the system should be focused on inherently safe design, detailed system specification, and feasibility analysis of the system. During the training phase, validation efforts could include analyzing the adequacy of the training data, verification of the training process, and evaluation of generalization capabilities. After training, the complete system should be validated

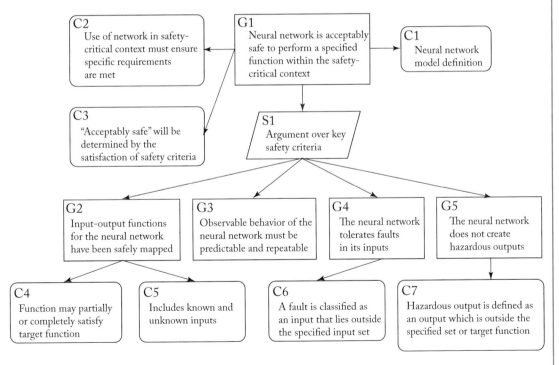

Figure 4.1: An example of a goal structuring notation for safety case argumentation of a deep neural network, where the boxes illustrate the goals, rounded boxes provide context for a goal, and rhomboids denote strategies to fulfill the goals. Adapted from [179].

for functional safety, through methods such as formal methods (if possible), black box testing, cross validation, and visualization. However, for online learners, validation should continue even after deployment. This will include methods from run-time monitoring and software safety cages which can ensure the safety of the system in real time. In summary, the discussion provided in this chapter indicates that the validation plan of a neural network system should include different techniques which have their own limitations and strengths to take advantage of their synergistic effects.

4.3   SUMMARY

This chapter focused on the safety validation of deep neural networks within autonomous vehicles. The difficulty of effective functional safety validation of deep neural networks is seen as a critical challenge for practical and deployable autonomous vehicles. The safety-critical nature of autonomous vehicles means that the safety of each sub-system must be validated to a high safety assurance. However, due to the complexity and opacity of these systems, there are currently no effective techniques to validate the safety of deep neural networks. This chapter discussed these issues in-depth, and reviewed on-going research attempting to solve such issues. Six classes of safety validation techniques were introduced, and the strengths and limitations of each class were discussed. Since each class has their own limitations, effective validation strategies will likely have to use techniques from all available classes.

CHAPTER 5

Concluding Remarks

In this book, a review of autonomous vehicle control through deep learning methods was completed. An introduction to deep learning and relevant concepts to later chapters were also presented. The review of control techniques was broken into three sections: lateral, longitudinal, and full vehicle control. The lateral control systems were shown to favor using supervised learning to predict steering angles from image inputs, while the dominant trend in longitudinal control was to use reinforcement learning from low dimensional ranging data. Meanwhile, in full vehicle control, a variety of approaches were seen. However, most recent approaches have mainly utilized vision as input, but a mix of reinforcement and supervised learning can still be seen. Important results have been achieved in singular driving tasks, such as lane keeping and collision avoidance. Research has also been carried out on learning more multi-modal behaviors, such as lane keeping systems which can also change lane or turn when required. While the works discussed in this book present impressive results, these systems are still far from human-level performance. However, research in this field is ongoing and quickly evolving, therefore improvements in the performance and complexity of deep learning systems can be expected in the future.

The main research challenges to deep neural network-driven autonomous vehicles were also identified and potential solutions were discussed. Computation is a challenge to deep learning systems due to the vast amount of training data required, especially when reinforcement learning is employed. Network architectures were identified as a challenge due to the reliance on heuristics in hyperparameter selection. Goal specification was discussed as a challenge in the context of reinforcement learning, as designing a reward function which adequately represents the desired behavior is difficult for complex tasks. Generalization is an issue in any neural network, but more so in complex operational environments where testing the performance for all possible scenarios is impossible. Verification and validation is challenge due to the need for accurate verification and validation of automotive systems and the difficulty of testing autonomous vehicles exhaustively in the real world. Finally, safety and interpretability was identified as the most critical challenge due to the safety-critical nature of autonomous driving and the opaque nature of deep neural networks.

The latter part of the book focused on the safety issues of using deep neural networks in safety-critical systems. Due to the lack of interpretability and complexity of these systems, validation of deep neural network-driven autonomous vehicles is challenging. Functional safety validation techniques for deep learning based autonomous vehicle systems were reviewed and analyzed. Several classes of techniques were outlined for safety validation of these systems, including formal methods, run-time monitoring, software safety cages, cross validation, visualization, and black box testing. Overall, the safety validation for these systems is extremely challenging due to the opaque, nonlinear, complex, and adaptive nature of the systems as well as the complexity and uncertainty of the operational environment. Recommendations were made to use a combination of multiple classes of techniques for their synergistic effects over all stages of the product lifecycle. Future work in the field will be required to improve the transparency of deep neural networks, thereby enabling the use of the traditionally powerful formal validation methods for practical deep neural networks. Additionally, meticulous analysis techniques for training sets would be beneficial to ensure adequate coverage of the expected operational environment. This would also require formal specification of the operational environment, which for an autonomous vehicle application would include characterizing all possible driving environments. Further work investigating effective defenses against adversarial attacks will also be needed before deployment of deep neural network-driven autonomous vehicles becomes feasible. Significant changes to the industry standards also need to be made to allow much of this to happen.

Bibliography

[1] Sebastian Thrun. Toward robotic cars. *Communications of the ACM*, 53(4):99, 2010. DOI: 10.1145/1721654.1721679 1

[2] Chris Urmson and William Whittaker. Self-driving cars and the urban challenge. *IEEE Intelligent Systems*, 23(2):66–68, 2008. DOI: 10.1109/mis.2008.34

[3] Umberto Montanaro, Shilp Dixit, Saber Fallah, Mehrdad Dianati, Alan Stevens, David Oxtoby, and Alexandros Mouzakitis. Towards connected autonomous driving: Review of use-cases. *Vehicle System Dynamics*, pages 1–36, 2018. DOI: 10.1080/00423114.2018.1492142 1

[4] World Health Organization. Global status report on road safety, 2018. DOI: 10.1136/ip.2009.023697 1

[5] Santokh Singh. Critical reasons for crashes investigated in the National Motor Vehicle Crash Causation Survey. *National Highway Traffic Safety Administration*, February 1–2, 2015. 1

[6] Thorsten Luettel, Michael Himmelsbach, and Hans Joachim Wuensche. Autonomous ground vehicles—concepts and a path to the future. *Proc. of the IEEE*, 100 (Special Centennial Issue):1831–1839, 2012. DOI: 10.1109/jproc.2012.2189803 1

[7] William Payre, Julien Cestac, and Patricia Delhomme. Intention to use a fully automated car: Attitudes and a priori acceptability. *Transportation Research Part F: Traffic Psychology and Behaviour*, 27(PB):252–263, 2014. DOI: 10.1016/j.trf.2014.04.009

[8] Philip Ross. Robot, you can drive my car. *IEEE Spectrum*, 51(6), 2014. DOI: 10.1109/mspec.2014.6821623

[9] Department for Transport. Research on the impacts of connected and autonomous vehicles (CAVs) on traffic flow, *Summary Report*, 2017. 1

[10] Kichun Jo, Junsoo Kim, Dongchul Kim, Chulhoon Jang, and Myoungho Sunwoo. Development of autonomous car—part I: Distributed system architecture and development process. *IEEE Transactions on Industrial Electronics*, 61(12):7131–7140, 2014. DOI: 10.1109/tie.2014.2321342 1

46 BIBLIOGRAPHY

[11] Charles Thorpe, Martial Herbert, Takeo Kanade, and S. Shafter. Toward autonomous driving: The CMU navlab. II. Architecture and systems. *IEEE Expert*, 6(4):44–52, 1991. DOI: 10.1109/64.85920 1

[12] Ernest D. Dickmanns and Alfred Zapp. Autonomous high speed road vehicle guidance by computer vision. *IFAC Proceedings Volumes*, 20(5):221–226, 1987. DOI: 10.1016/s1474-6670(17)55320-3 1

[13] Sebastian Thrun, Mike Montemerlo, Hendrik Dahlkamp, David Stavens, Andrei Aron, James Diebel, Philip Fong, John Gale, Morgan Halpenny, Gabriel Hoffmann, Kenny Lau, Celia Oakley, Mark Palatucci, Vaughan Pratt, Pascal Stang, Sven Strohband, Cedric Dupont, Lars-Erik Jendrossek, Christian Koelen, Charles Markey, Carlo Rummel, Joe van Niekerk, Eric Jensen, Philippe Alessandrini, Gary Bradski, Bob Davies, Scott Ettinger, Adrian Kaehler, Ara Nefian, Pamela Mahoney. Stanley: The robot that won the DARPA grand challenge. *Journal of Field Robotics*, 23(9):661–692, 2006. DOI: 10.1007/978-3-540-73429-1_1 1

[14] Martin Buehler, Karl Iagnemma, and Sanjiv Singh. *The DARPA Urban Challenge: Autonomous Vehicles in City Traffic*, vol. 56, Springer, 2009. DOI: 10.1007/978-3-642-03991-1 1

[15] SAE International. Taxonomy and definitions for terms related to driving automation systems for on-road motor vehicles. *SAE International, (J3016)*, 2018. DOI: 10.4271/J3016_201609 1

[16] Aldo Sorniotti, Phil Barber, and Stefano De Pinto. Path tracking for automated driving: A tutorial on control system formulations and ongoing research. In *Automated Driving*, pages 71–140, Springer, 2017. DOI: 10.1007/978-3-319-31895-0_5 2

[17] Tuan Le-Anh and M. B. M. De Koster. A review of design and control of automated guided vehicle systems. *European Journal of Operational Research*, 171(1):1–23, 2006. DOI: 10.1016/j.ejor.2005.01.036 2

[18] Brian Paden, Michal Čáp, Sze Zheng Yong, Dmitry Yershov, and Emilio Frazzoli. A survey of motion planning and control techniques for self-driving urban vehicles. *IEEE Transactions on Intelligent Vehicles*, 1(1):33–55, 2016. DOI: 10.1109/tiv.2016.2578706

[19] Michel Pasquier, Chai Quek, and Mary Toh. Fuzzylot: A novel self-organising fuzzy-neural rule-based pilot system for automated vehicles. *Neural Networks*, 14(8):1099–1112, 2001. DOI: 10.1016/s0893-6080(01)00048-x 2

[20] Markus Kuderer, Shilpa Gulati, and Wolfram Burgard. Learning driving styles for autonomous vehicles from demonstration. *Proc. of the IEEE International Conference on*

Robotics and Automation, pages 2641–2646, June 2015. DOI: 10.1109/icra.2015.7139555
3

[21] David Silver, J. Andrew Bagnell, and Anthony Stentz. Learning autonomous driving styles and maneuvers from expert demonstration. In *Experimental Robotics*, pages 371–386, Springer, Heidelberg, 2013. DOI: 10.1007/978-3-319-00065-7_26 3

[22] Dongbin Zhao, Bin Wang, and Derong Liu. A supervised actor-critic approach for adaptive cruise control. *Soft Computing*, 17(11):2089–2099, 2013. DOI: 10.1007/s00500-013-1110-y 3, 20, 21

[23] C. Desjardins and B. Chaib-draa. Cooperative adaptive cruise control: A reinforcement learning approach. *IEEE Transactions on Intelligent Transportation Systems*, 12(4):1248–1260, 2011. DOI: 10.1109/tits.2011.2157145 3, 18, 21

[24] Alex Krizhevsky, Ilya Sutskever, and Geoffrey E. Hinton. Imagenet classification with deep convolutional neural networks. In *Advances in Neural Information Processing Systems*, pages 1097–1105, 2012. DOI: 10.1145/3065386 3, 5, 9

[25] Geoffrey Hinton, Li Deng, Dong Yu, George E. Dahl, Abdel-Rahman Mohamed, Navdeep Jaitly, Andrew Senior, Vincent Vanhoucke, Patrick Nguyen, Tara N. Sainath, Brian Kingsbury. Deep neural networks for acoustic modeling in speech recognition: The shared views of four research groups. *IEEE Signal Processing Magazine*, 29(6):82–97, 2012. DOI: 10.1109/msp.2012.2205597 5

[26] Ilya Sutskever, Oriol Vinyals, and Quoc V. Le. Sequence to sequence learning with neural networks. In *Advances in Neural Information Processing Systems*, pages 3104–3112, 2014. 3, 5

[27] Wilko Schwarting, Javier Alonso-Mora, and Daniela Rus. Planning and decision-making for autonomous vehicles. *Annual Review of Control, Robotics, and Autonomous Systems*, 1:187–210, 2018. DOI: 10.1146/annurev-control-060117-105157 3

[28] Thi Thoa Mac, Cosmin Copot, Duc Trung Tran, and Robin De Keyser. Heuristic approaches in robot path planning: A survey. *Robotics and Autonomous Systems*, 86:13–28, 2016. DOI: 10.1016/j.robot.2016.08.001

[29] Sandor M. Veres, Levente Molnar, Nick K. Lincoln, and Colin P. Morice. Autonomous vehicle control systems—a review of decision making. *Proc. of the Institution of Mechanical Engineers, Part I: Journal of Systems and Control Engineering*, 225(2):155–195, 2011. DOI: 10.1177/2041304110394727 3

[30] Hao Ye and Geoffrey Ye Li. Deep reinforcement learning for resource allocation in v2v communications. In *IEEE International Conference on Communications (ICC)*, pages 1–6, 2018. DOI: 10.1109/icc.2018.8422586 3

[31] Xianfu Chen, Celimuge Wu, Honggang Zhang, Yan Zhang, Mehdi Bennis, and Heli Vuojala. Decentralized deep reinforcement learning for delay-power tradeoff in vehicular communications. *ArXiv Preprint ArXiv:1906.00625*, 2019. 3

[32] Hao Zhu, Ka-Veng Yuen, Lyudmila Mihaylova, and Henry Leung. Overview of environment perception for intelligent vehicles. *IEEE Transactions on Intelligent Transportation Systems*, 18(10):2584–2601, 2017. DOI: 10.1109/tits.2017.2658662 3

[33] Jessica Van Brummelen, Marie O'Brien, Dominique Gruyer, and Homayoun Najjaran. Autonomous vehicle perception: The technology of today and tomorrow. *Transportation Research Part C: Emerging Technologies*, vol. 89, pages 384–406, April 2018. DOI: 10.1016/j.trc.2018.02.012

[34] Joel Janai, Fatma Güney, Aseem Behl, and Andreas Geiger. Computer vision for autonomous vehicles: Problems, datasets and state-of-the-art. *ArXiv Preprint ArXiv:1704.05519*, 2017. 3

[35] Stephanie Lowry, Niko Sünderhauf, Paul Newman, John J. Leonard, David Cox, Peter Corke, and Michael J. Milford. Visual place recognition: A survey. *IEEE Transactions on Robotics*, 32(1):1–19, 2016. DOI: 10.1109/tro.2015.2496823 3

[36] Kishore Reddy Konda and Roland Memisevic. Learning visual odometry with a convolutional network. In *VISAPP (1)*, pages 486–490, 2015. DOI: 10.5220/0005299304860490

[37] Sampo Kuutti, Saber Fallah, Konstantinos Katsaros, Mehrdad Dianati, Francis Mccullough, and Alexandros Mouzakitis. A survey of the state-of-the-art localization techniques and their potentials for autonomous vehicle applications. *IEEE Internet of Things Journal*, 5(2):829–846, 2018. DOI: 10.1109/jiot.2018.2812300 3

[38] Volodymyr Mnih, Koray Kavukcuoglu, David Silver, Andrei A. Rusu, Joel Veness, Marc G. Bellemare, Alex Graves, Martin Riedmiller, Andreas K. Fidjeland, Georg Ostrovski, et al. Human-level control through deep reinforcement learning. *Nature*, 518(7540):529, 2015. DOI: 10.1038/nature14236 3

[39] Itamar Arel, Derek C. Rose, and Thomas P. Karnowski. Deep machine learning-a new frontier in artificial intelligence research [research frontier]. *IEEE Computational Intelligence Magazine*, 5(4):13–18, 2010. DOI: 10.1109/mci.2010.938364

[40] Jun Tani, Masato Ito, and Yuuya Sugita. Self-organization of distributedly represented multiple behavior schemata in a mirror system: Reviews of robot experiments using RN-NPB. *Neural Networks*, 17(8–9):1273–1289, 2004. DOI: 10.1016/j.neunet.2004.05.007

[41] Yann Lecun, Yoshua Bengio, and Geoffrey Hinton. Deep learning. *Nature*, 521(7553):436–444, 2015. DOI: 10.1038/nature14539

[42] Jürgen Schmidhuber. Deep learning in neural networks: An overview. *Neural Networks*, 61:85–117, 2015. DOI: 10.1016/j.neunet.2014.09.003 3

[43] Sergey Levine, Chelsea Finn, Trevor Darrell, and Pieter Abbeel. End-to-end training of deep visuomotor policies. *The Journal of Machine Learning Research*, 17(1):1334–1373, 2016. 3, 18

[44] Sergey Levine, Peter Pastor, Alex Krizhevsky, and Deirdre Quillen. Learning hand-eye coordination for robotic grasping with large-scale data collection. In *International Symposium on Experimental Robotics*, pages 173–184, Springer, 2016. DOI: 10.1007/978-3-319-50115-4_16 18

[45] Viktor Rausch, Andreas Hansen, Eugen Solowjow, Edwin Kreuzer, and J. Karl Hedrick. Learning a deep neural net policy for end-to-end control of autonomous vehicles, *American Control Conference (ACC)*, pages 4914–4919, 2017. DOI: 10.23919/acc.2017.7963716 3, 16, 17

[46] Yoshua Bengio, Nicolas Boulanger-Lewandowski, and Razvan Pascanu. Advances in optimizing recurrent networks. In *Acoustics, Speech and Signal Processing (ICASSP), IEEE International Conference on*, pages 8624–8628, 2013. DOI: 10.1109/icassp.2013.6639349 7

[47] Razvan Pascanu, Tomas Mikolov, and Yoshua Bengio. On the difficulty of training recurrent neural networks. In *International Conference on Machine Learning*, pages 1310–1318, 2013.

[48] James Martens and Ilya Sutskever. Learning recurrent neural networks with hessian-free optimization. In *Proc. of the 28th International Conference on Machine Learning (ICML-11)*, pages 1033–1040, Citeseer, 2011.

[49] Ilya Sutskever. Training recurrent neural networks. Ph.D. thesis, University of Toronto, Toronto, Ontario, Canada, 2013.

[50] Ilya Sutskever, James Martens, George Dahl, and Geoffrey Hinton. On the importance of initialization and momentum in deep learning. In *International Conference on Machine Learning*, pages 1139–1147, 2013. 7

[51] Sepp Hochreiter and Jürgen Schmidhuber. Long short-term memory. *Neural Computation*, 9(8):1735–1780, 1997. DOI: 10.1162/neco.1997.9.8.1735 7

[52] Kyunghyun Cho, Bart Van Merriënboer, Caglar Gulcehre, Dzmitry Bahdanau, Fethi Bougares, Holger Schwenk, and Yoshua Bengio. Learning phrase representations using RNN encoder-decoder for statistical machine translation. *ArXiv Preprint ArXiv:1406.1078*, 2014. DOI: 10.3115/v1/d14-1179 7

[53] Christian Szegedy, Wei Liu, Yangqing Jia, Pierre Sermanet, Scott Reed, Dragomir Anguelov, Dumitru Erhan, Vincent Vanhoucke, and Andrew Rabinovich. Going deeper with convolutions. In *Proc. of the IEEE Conference on Computer Vision and Pattern Recognition*, pages 1–9, 2015. DOI: 10.1109/cvpr.2015.7298594 9

[54] Sergey Ioffe and Christian Szegedy. Batch normalization: Accelerating deep network training by reducing internal covariate shift. *ArXiv Preprint ArXiv:1502.03167*, 2015. 10

[55] Quentin J. M. Huys, Anthony Cruickshank, and Peggy Seriès. Reward-based learning, model-based and model-free. In *Encyclopedia of Computational Neuroscience*, pages 2634–2641, Springer, 2015. DOI: 10.1007/978-1-4614-6675-8_674 11

[56] R. S. Sutton and A. G. Barto. *Reinforcement Learning: An Introduction*, vol. 9, MIT Press, Cambridge, MA, 1998. DOI: 10.1109/tnn.1998.712192 11, 13, 20

[57] Marco Wiering and Martijn Van Otterlo. Reinforcement learning. *Adaptation, Learning, and Optimization*, 12:51, 2012. DOI: 10.1007/978-3-642-27645-3

[58] Kai Arulkumaran, Marc Peter Deisenroth, Miles Brundage, and Anil Anthony Bharath. Deep reinforcement learning: A brief survey. *IEEE Signal Processing Magazine*, 34(6):26–38, 2017. DOI: 10.1109/msp.2017.2743240 11

[59] Vijay R. Konda and John N. Tsitsiklis. Onactor-critic algorithms. *SIAM Journal on Control and Optimization*, 42(4):1143–1166, 2003. DOI: 10.1137/s0363012901385691 12

[60] Christopher John Cornish Hellaby Watkins. Learning from delayed rewards. Ph.D. thesis, King's College, Cambridge, 1989. 12

[61] Christopher J. C. H. Watkins and Peter Dayan. Q-learning. *Machine Learning*, 8(3–4):279–292, 1992. DOI: 10.1007/bf00992698

[62] Steven J. Bradtke, B. Erik Ydstie, and Andrew G. Barto. Adaptive linear quadratic control using policy iteration. In *Proc. of the American Control Conference*, vol. 3, pages 3475–3475, Citeseer, 1994. DOI: 10.1109/acc.1994.735224 12

[63] Gavin A. Rummery and Mahesan Niranjan. *On-Line Q-Learning Using Connectionist Systems*, vol. 37, University of Cambridge, Department of Engineering Cambridge, UK, 1994. 12

[64] Ivo Grondman, Lucian Busoniu, Gabriel A. D. Lopes, and Robert Babuska. A survey of actor-critic reinforcement learning: Standard and natural policy gradients. *IEEE Transactions on Systems, Man, and Cybernetics, Part C (Applications and Reviews)*, 42(6):1291–1307, 2012. DOI: 10.1109/tsmcc.2012.2218595 12

[65] Leemon Baird. Residual algorithms: Reinforcement learning with function approximation. In *Machine Learning Proceedings*, pages 30–37, Elsevier, 1995. DOI: 10.1016/b978-1-55860-377-6.50013-x

[66] Geoffrey J. Gordon. Stable function approximation in dynamic programming. In *Machine Learning Proceedings*, pages 261–268, Elsevier, 1995. DOI: 10.1016/b978-1-55860-377-6.50040-2

[67] John N. Tsitsiklis and Benjamin Van Roy. Feature-based methods for large scale dynamic programming. *Machine Learning*, 22(1–3):59–94, 1996. DOI: 10.1007/978-0-585-33656-5_5 12

[68] Ronald J. Williams. *Reinforcement-Learning Connectionist Systems*. College of Computer Science, Northeastern University, 1987. 12

[69] R. Williams. A class of gradient-estimation algorithms for reinforcement learning in neural networks. In *Proc. of the International Conference on Neural Networks*, pages II–601, 1987.

[70] Ronald J. Williams. Simple statistical gradient-following algorithms for connectionist reinforcement learning. *Machine Learning*, 8(3–4):229–256, 1992. DOI: 10.1007/978-1-4615-3618-5_2 12

[71] Richard S. Sutton, David A. McAllester, Satinder P. Singh, and Yishay Mansour. Policy gradient methods for reinforcement learning with function approximation. In *Advances in Neural Information Processing Systems*, pages 1057–1063, 2000. 12

[72] Martin Riedmiller, Jan Peters, and Stefan Schaal. Evaluation of policy gradient methods and variants on the cart-pole benchmark. In *Approximate Dynamic Programming and Reinforcement Learning, (ADPRL). IEEE International Symposium on*, pages 254–261, 2007. DOI: 10.1109/adprl.2007.368196 12

[73] Volodymyr Mnih, Adria Puigdomenech Badia, Mehdi Mirza, Alex Graves, Timothy Lillicrap, Tim Harley, David Silver, and Koray Kavukcuoglu. Asynchronous methods for deep reinforcement learning. In *International Conference on Machine Learning*, pages 1928–1937, 2016. 12

[74] Timothy P. Lillicrap, Jonathan J. Hunt, Alexander Pritzel, Nicolas Heess, Tom Erez, Yuval Tassa, David Silver, and Daan Wierstra. Continuous control with deep reinforcement learning. *ArXiv Preprint ArXiv:1509.02971*, 2015. 12

[75] M. Barto and Michael T. Rosenstein. J. 4 supervised actor-critic reinforcement learning. *Handbook of Learning and Approximate Dynamic Programming*, 2:359, 2004. 12

[76] Vijay R. Konda and John N. Tsitsiklis. Actor-critic algorithms. In *Advances in Neural Information Processing Systems*, pages 1008–1014, 2000. DOI: 10.1137/s0363012901385691 12

[77] Michael Nielsen. *Neural Networks and Deep Learning*. Determination Press, 2015. 13

[78] Ian Goodfellow, Yoshua Bengio, and Aaron Courville. *Deep Learning*, MIT Press, 2016. 13

[79] Marc Peter Deisenroth, A. Aldo Faisal, and Cheng Soon Ong. *Mathematics for Machine Learning*. Cambridge University Press, 2019. 13

[80] Trevor Hastie, Robert Tibshirani, and Jerome Friedman. *The Elements of Statistical Learning: Data Mining, Inference, and Prediction*, Springer Series in Statistics, Springer, New York, 2009. DOI: 10.1007/978-0-387-21606-5 13

[81] Francois Chollet. *Deep Learning with Python*. Manning Publications Co., 2017. 13

[82] Aurélien Géron. *Hands-On Machine Learning with Scikit-Learn and TensorFlow: Concepts, Tools, and Techniques to Build Intelligent Systems*. O'Reilly Media, Inc., 2017. 13

[83] Dean A. Pomerleau. Alvinn: An autonomous land vehicle in a neural network. *Advances in Neural Information Processing Systems 1*, pages 305–313, 1989. 15, 17, 24

[84] D. Pomerleau. Neural network vision for robot driving. *Intelligent Unmanned Ground Vehicles*, pages 1–22, 1997. DOI: 10.1007/978-1-4615-6325-9_4 15, 17

[85] Gening Yu and I. K. Sethi. Road-following with continuous learning. In *Intelligent Vehicles Symposium. Proceedings of the*, Detroit, MI, 1995. DOI: 10.1109/ivs.1995.528317 16, 17

[86] Urs Muller, Jan Ben, Eric Cosatto, Beat Flepp, and Yann L. Cun. Off-road obstacle avoidance through end-to-end learning. In *Advances in Neural Information Processing Systems*, pages 739–746, 2006. 16, 17

[87] Mariusz Bojarski, Davide Del Testa, Daniel Dworakowski, Bernhard Firner, Beat Flepp, Prasoon Goyal, Lawrence D. Jackel, Mathew Monfort, Urs Muller, Jiakai Zhang, et al. End to end learning for self-driving cars. *ArXiv Preprint ArXiv:1604.07316*, 2016. 16, 17, 37

[88] Hesham M. Eraqi, Mohamed N. Moustafa, and Jens Honer. End-to-end deep learning for steering autonomous vehicles considering temporal dependencies. *ArXiv Preprint ArXiv:1710.03804*, 2017. 16, 17

[89] Ardalan Vahidi and Azim Eskandarian. Research advances in intelligent collision avoidance and adaptive cruise control. *IEEE Transactions on Intelligent Transportation Systems*, 4(3):143–153, 2003. DOI: 10.1109/tits.2003.821292 18

[90] Seungwuk Moon, Ilki Moon, and Kyongsu Yi. Design, tuning, and evaluation of a full-range adaptive cruise control system with collision avoidance. *Control Engineering Practice*, 17(4):442–455, 2009. DOI: 10.1016/j.conengprac.2008.09.006 18

[91] Qi Sun. Cooperative adaptive cruise control performance analysis. Ph.D. thesis, Ecole Centrale de Lille, 2016. 18

[92] Hassan K. Khalil. *Nonlinear Systems*, 2(5):5–1, Prentice Hall, NJ, 1996. 18

[93] Dan Wang and Jie Huang. Neural network-based adaptive dynamic surface control for a class of uncertain nonlinear systems in strict-feedback form. *IEEE Transactions on Neural Networks*, 16(1):195–202, 2005. DOI: 10.1109/tnn.2004.839354 18

[94] Xiaohui Dai, Chi-Kwong Li, and Ahmad B. Rad. An approach to tune fuzzy controllers based on reinforcement learning for autonomous vehicle control. *IEEE Transactions on Intelligent Transportation Systems*, 6(3):285–293, 2005. DOI: 10.1109/fuzz.2003.1209417 18, 21

[95] Zhenhua Huang, Xin Xu, Haibo He, Jun Tan, and Zhenping Sun. Parameterized batch reinforcement learning for longitudinal control of autonomous land vehicles. *IEEE Transactions on Systems, Man, and Cybernetics: Systems*, pages 1–12, 2017. DOI: 10.1109/tsmc.2017.2712561 19, 21

[96] Xin Chen, Yong Zhai, Chao Lu, Jianwei Gong, and Gang Wang. A learning model for personalized adaptive cruise control. In *Intelligent Vehicles Symposium (IV), IEEE*, pages 379–384, 2017. DOI: 10.1109/ivs.2017.7995748 19, 21

[97] Dongbin Zhao, Zhongpu Xia, and Qichao Zhang. Model-free optimal control based intelligent cruise control with hardware-in-the-loop demonstration (research frontier). *IEEE Computational Intelligence Magazine*, 12(2):56–69, 2017. DOI: 10.1109/mci.2017.2670463 19, 21

[98] Hyunmin Chae, Chang Mook Kang, ByeoungDo Kim, Jaekyum Kim, Chung Choo Chung, and Jun Won Choi. Autonomous braking system via deep reinforcement learning. In *IEEE 20th International Conference on Intelligent Transportation Systems (ITSC)*, pages 1–6, 2017. DOI: 10.1109/itsc.2017.8317839 19, 21, 24

[99] Euro NCAP. European new car assessment programme: Test protocol—AEB VRU systems, 2015. 19

[100] Leslie Pack Kaelbling, Michael L. Littman, and Andrew W. Moore. Reinforcement learning: A survey. *Journal of Artificial Intelligence Research*, 4:237–285, 1996. DOI: 10.1613/jair.301 20

[101] Dongbin Zhao, Zhaohui Hu, Zhongpu Xia, Cesare Alippi, Yuanheng Zhu, and Ding Wang. Full-range adaptive cruise control based on supervised adaptive dynamic programming. *Neurocomputing*, 125:57–67, February 2014. DOI: 10.1016/j.neucom.2012.09.034 20

[102] Bin Wang, Dongbin Zhao, Chengdong Li, and Yujie Dai. Design and implementation of an adaptive cruise control system based on supervised actor-critic learning. *5th International Conference on Information Science and Technology (ICIST)*, pages 243–248, 2015. DOI: 10.1109/icist.2015.7288976 20

[103] Rui Zheng, Chunming Liu, and Qi Guo. A decision-making method for autonomous vehicles based on simulation and reinforcement learning. *International Conference on Machine Learning and Cybernetics*, pages 362–369, 2013. DOI: 10.1109/icmlc.2013.6890495 20

[104] Shai Shalev-Shwartz, Nir Ben-Zrihem, Aviad Cohen, and Amnon Shashua. Long-term planning by short-term prediction. *ArXiv Preprint ArXiv:1602.01580*, 2016. 22, 26

[105] Wei Xia, Huiyun Li, and Baopu Li. A control strategy of autonomous vehicles based on deep reinforcement learning. In *Computational Intelligence and Design (ISCID), 9th International Symposium on*, vol. 2, pages 198–201, IEEE, 2016. DOI: 10.1109/iscid.2016.2054 22, 26

[106] Ahmad El Sallab, Mohammed Abdou, Etienne Perot, and Senthil Yogamani. End-to-end deep reinforcement learning for lane keeping assist. *ArXiv Preprint ArXiv:1612.04340*, 2016. 22, 26

[107] Jiakai Zhang and Kyunghyun Cho. Query-efficient imitation learning for end-to-end autonomous driving. *ArXiv Preprint ArXiv:1605.06450*, 2016. 22, 26

[108] Stéphane Ross, Geoffrey Gordon, and Drew Bagnell. A reduction of imitation learning and structured prediction to no-regret online learning. In *Proc. of the 14th International Conference on Artificial Intelligence and Statistics*, pages 627–635, 2011. 22

[109] Yunpeng Pan, Ching-An Cheng, Kamil Saigol, Keuntaek Lee, Xinyan Yan, Evangelos Theodorou, and Byron Boots. Agile autonomous driving using end-to-end deep imitation learning. *Proc. of Robotics: Science and Systems*, Pittsburgh, PA, 2018. DOI: 10.15607/rss.2018.xiv.056 23, 26

[110] Dequan Wang, Coline Devin, Qi-Zhi Cai, Fisher Yu, and Trevor Darrell. Deep object centric policies for autonomous driving. *ArXiv Preprint ArXiv:1811.05432*, 2018. 23, 26

[111] Horia Porav and Paul Newman. Imminent collision mitigation with reinforcement learning and vision. In *21st International Conference on Intelligent Transportation Systems (ITSC)*, pages 958–964, IEEE, 2018. DOI: 10.1109/itsc.2018.8569222 23, 26

[112] Simon Hecker, Dengxin Dai, and Luc Van Gool. End-to-end learning of driving models with surround-view cameras and route planners. In *Proc. of the European Conference on Computer Vision (ECCV)*, pages 435–453, 2018. DOI: 10.1007/978-3-030-01234-2_27 24, 26

[113] Felipe Codevilla, Matthias Müller, Antonio López, Vladlen Koltun, and Alexey Dosovitskiy. End-to-end driving via conditional imitation learning. In *IEEE International Conference on Robotics and Automation (ICRA)*, pages 1–9, 2018. DOI: 10.1109/icra.2018.8460487 24, 26

[114] Mayank Bansal, Alex Krizhevsky, and Abhijit Ogale. Chauffeurnet: Learning to drive by imitating the best and synthesizing the worst. *ArXiv Preprint ArXiv:1812.03079*, 2018. 24, 26, 28

[115] Markus Maurer, J. Christian Gerdes, Barbara Lenz, and Hermann Winner. *Autonomous Driving*. Springer, Heidelberg, Berlin, 2016. DOI: 10.1007/978-3-662-48847-8 25

[116] European Commission. Cooperative intelligent transportation systems—research theme analysis report, 2016. https://trimis.ec.europa.eu/sites/default/files/TRIP _C-ITS_Report.pdf

[117] Saeed Asadi Bagloee, Madjid Tavana, Mohsen Asadi, and Tracey Oliver. Autonomous vehicles: Challenges, opportunities, and future implications for transportation policies. *Journal of Modern Transportation*, 24(4):284–303, 2016. DOI: 10.1007/s40534-016-0117-3

[118] HERE Technologies. Consumer acceptance of autonomous vehicles, 2017. https://www.here.com/sites/g/files/odxslz166/files/2018-11/Consumer _Acceptance_of_Autonomous_Vehicles_white_paper_0.pdf

[119] Leopold Bosankic. How consumers' perception of autonomous cars will influence their adoption, 2017. https://medium.com/@leo_pold_b/how-consumers-percepti on-of-autonomous-cars-will-influence-their-adoption-ba99e3f64e9a

[120] Hillary Abraham, Bryan Reimer, Bobbie Seppelt, Craig Fitzgerald, Bruce Mehler, and Joseph F. Coughlin. Consumer interest in automation: Preliminary observations exploring a year's change, 2017. http://agelab.mit.edu/sites/default/files/MIT%20-%20N EMPA%20White%20Paper%20FINAL.pdf 25

[121] Yoshua Bengio. Deep learning of representations: Looking forward. In *International Conference on Statistical Language and Speech Processing*, pages 1–37, Springer, 2013. DOI: 10.1007/978-3-642-39593-2_1 25

[122] Maryam M. Najafabadi, Flavio Villanustre, Taghi M. Khoshgoftaar, Naeem Seliya, Randall Wald, and Edin Muharemagic. Deep learning applications and challenges in big data analytics. *Journal of Big Data*, 2(1):1, 2015. DOI: 10.1186/s40537-014-0007-7

[123] Tal Ben-Nun and Torsten Hoefler. Demystifying parallel and distributed deep learning: An in-depth concurrency analysis. *ArXiv Preprint ArXiv:1802.09941*, 2018. 25

[124] Ziyu Wang, Victor Bapst, Nicolas Heess, Volodymyr Mnih, Remi Munos, Koray Kavukcuoglu, and Nando de Freitas. Sample efficient actor-critic with experience replay. *ArXiv Preprint ArXiv:1611.01224*, 2016. 25

[125] James Bergstra and Yoshua Bengio. Random search for hyper-parameter optimization. *Journal of Machine Learning Research*, 13:281–305, 2012. 27

[126] Colin Raffel. Neural network hyperparameters, 2015. http://colinraffel.com/wiki/neural_network_hyperparameters 27

[127] Yoshua Bengio. Practical recommendations for gradient-based training of deep architectures. In *Neural Networks: Tricks of the Trade*, pages 437–478, Springer, 2012. DOI: 10.1007/978-3-642-35289-8_26 27

[128] Yann LeCun, Léon Bottou, Genevieve B. Orr, and Klaus-Robert Müller. Efficient backprop. In *Neural Networks: Tricks of the Trade*, pages 9–50, Springer, 1998. DOI: 10.1007/3-540-49430-8_2 27

[129] Y. LeCun. Generalization and network design strategies. *Connectionism in Perspective*, pages 143–155, 1989. 27

[130] Nelson Morgan and Hervé Bourlard. Generalization and parameter estimation in feedforward nets: Some experiments. *Advances in Neural Information Processing Systems*, pages 630–637, 1989. 27

[131] Thomas Elsken, Jan Hendrik Metzen, and Frank Hutter. Neural architecture search: A survey. *Journal of Machine Learning Research*, 20(55):1–21, 2019. DOI: 10.1007/978-3-030-05318-5_3 27

[132] Matthias Feurer and Frank Hutter. Hyperparameter optimization. In *Automatic Machine Learning: Methods, Systems, Challenges*, pages 3–38, Springer, 2018. http://automl.org/book DOI: 10.1007/978-3-030-05318-5_1

[133] James Bergstra, Rémi Bardenet, Yoshua Bengio, and Balázs Kégl. Algorithms for hyperparameter optimization. In *Advances in Neural Information Processing Systems (NIPS)*, pages 2546–2554, 2011.

[134] Manoj Kumar, George E. Dahl, Vijay Vasudevan, and Mohammad Norouzi. Parallel architecture and hyperparameter search via successive halving and classification. *ArXiv Preprint ArXiv:1805.10255*, 2018.

[135] Tatsunori B. Hashimoto, Steve Yadlowsky, and John C. Duchi. Derivative free optimization via repeated classification. *ArXiv Preprint ArXiv:1804.03761*, 2018. 27

[136] Andrew Y. Ng, Daishi Harada, and Stuart Russell. Policy invariance under reward transformations: Theory and application to reward shaping. *16th International Conference on Machine Learning*, 3:278–287, 1999. 27

[137] A. D. Laud. Theory and application of reward shaping in reinforcement learning. Ph.D. thesis, University of Illinois, 2004. 27

[138] Kober, Jens J., Bagnell, Andrew, Peters, Jan. Reinforcement learning in robotics: A survey. *International Journal of Robotics Research*, 32(11):1238–1274, 2013. DOI: 10.1177/0278364913495721 27

[139] Dario Amodei, Chris Olah, Jacob Steinhardt, Paul Christiano, John Schulman, and Dan Mané. Concrete problems in AI safety. *ArXiv Preprint ArXiv:1606.06565*, 2016. 27, 29

[140] Nidhi Kalra and Susan M. Paddock. Driving to safety: How many miles of driving would it take to demonstrate autonomous vehicle reliability? *Transportation Research Part A: Policy and Practice*, 94:182–193, 2016. DOI: 10.1016/j.tra.2016.09.010 27, 31

[141] Simon Burton, Lydia Gauerhof, and Christian Heinzemann. Making the case for safety of machine learning in highly automated driving. In *International Conference on Computer Safety, Reliability, and Security*, pages 5–16, Springer, 2017. DOI: 10.1007/978-3-319-66284-8_1 27, 31, 38, 40

[142] Andrew Y. Ng. Feature selection, l 1 vs. l 2 regularization, and rotational invariance. In *Proc. of the 21st International Conference on Machine Learning*, page 78, ACM, 2004. DOI: 10.1145/1015330.1015435 28

[143] Paul Merolla, Rathinakumar Appuswamy, John Arthur, Steve K. Esser, and Dharmendra Modha. Deep neural networks are robust to weight binarization and other non-linear distortions. *ArXiv Preprint ArXiv:1606.01981*, 2016.

[144] Matthieu Courbariaux, Yoshua Bengio, and Jean-Pierre David. Binaryconnect: Training deep neural networks with binary weights during propagations. In *Advances in Neural Information Processing Systems*, pages 3123–3131, 2015.

[145] Nitish Srivastava, Geoffrey Hinton, Alex Krizhevsky, Ilya Sutskever, and Ruslan Salakhutdinov. Dropout: A simple way to prevent neural networks from overfitting. *Journal of Machine Learning Research*, 15:1929–1958, 2014.

[146] Geoffrey E. Hinton, Nitish Srivastava, Alex Krizhevsky, Ilya Sutskever, and Ruslan R. Salakhutdinov. Improving neural networks by preventing co-adaptation of feature detectors. *ArXiv Preprint ArXiv:1207.0580*, 2012. 28

[147] Luke Ng, Christopher M. Clark, and Jan P. Huissoon. Reinforcement learning of adaptive longitudinal vehicle control for dynamic collaborative driving. In *IEEE Intelligent Vehicles Symposium, Proceedings*, pages 907–912, 2008. DOI: 10.1109/ivs.2008.4621222 28

[148] Josh Tobin, Rachel Fong, Alex Ray, Jonas Schneider, Wojciech Zaremba, and Pieter Abbeel. Domain randomization for transferring deep neural networks from simulation to the real world. In *Intelligent Robots and Systems (IROS), IEEE/RSJ International Conference on*, pages 23–30, 2017. DOI: 10.1109/iros.2017.8202133 28

[149] Fereshteh Sadeghi and Sergey Levine. Cad2rl: Real single-image flight without a single real image. *ArXiv Preprint ArXiv:1611.04201*, 2016. DOI: 10.15607/rss.2017.xiii.034

[150] Stephen James, Andrew J. Davison, and Edward Johns. Transferring end-to-end visuomotor control from simulation to real world for a multi-stage task. *ArXiv Preprint ArXiv:1707.02267*, 2017. 28

[151] Andrei A. Rusu, Matej Vecerik, Thomas Rothörl, Nicolas Heess, Razvan Pascanu, and Raia Hadsell. Sim-to-real robot learning from pixels with progressive nets. *ArXiv Preprint ArXiv:1610.04286*, 2016. 28

[152] Ali Ghadirzadeh, Atsuto Maki, Danica Kragic, and Mårten Björkman. Deep predictive policy training using reinforcement learning. In *IEEE/RSJ International Conference on Intelligent Robots and Systems (IROS)*, pages 2351–2358, 2017. DOI: 10.1109/iros.2017.8206046

[153] Abhishek Gupta, Coline Devin, YuXuan Liu, Pieter Abbeel, and Sergey Levine. Learning invariant feature spaces to transfer skills with reinforcement learning. *ArXiv Preprint ArXiv:1703.02949*, 2017.

[154] Jason Yosinski, Jeff Clune, Yoshua Bengio, and Hod Lipson. How transferable are features in deep neural networks? In *Advances in Neural Information Processing Systems*, pages 3320–3328, 2014. 28

[155] Pieter Abbeel, Morgan Quigley, and Andrew Y. Ng. Using inaccurate models in reinforcement learning. In *Proc. of the 23rd International Conference on Machine Learning*, pages 1–8, ACM, 2006. DOI: 10.1145/1143844.1143845 28

[156] Mark Cutler, Thomas J. Walsh, and Jonathan P. How. Reinforcement learning with multi-fidelity simulators. In *IEEE International Conference on Robotics and Automation (ICRA)*, pages 3888–3895, 2014. DOI: 10.1109/icra.2014.6907423

[157] Jur Van Den Berg, Stephen Miller, Daniel Duckworth, Humphrey Hu, Andrew Wan, Xiao-Yu Fu, Ken Goldberg, and Pieter Abbeel. Superhuman performance of surgical tasks by robots using iterative learning from human-guided demonstrations. In *IEEE International Conference on Robotics and Automation*, pages 2074–2081, 2010. DOI: 10.1109/robot.2010.5509621 28

[158] T. Hester, M. Vecerik, O. Pietquin, M. Lanctot, T. Schaul, B. Piot, D. Horgan, J. Quan, A. Sendonaris, G. Dulac-Arnold, et al. Learning from demonstrations for real world reinforcement learning. ArXiv, 2017. 28

[159] Davide Castelvecchi. Can we open the black box of AI? *Nature News*, 538(7623):20, 2016. DOI: 10.1038/538020a 28, 31

[160] Perry Van Wesel and Alwyn E. Goodloe. Challenges in the verification of reinforcement learning algorithms. Technical report, NASA, 2017. 29, 39

[161] International Organization for Standardization. ISO 26262: Road vehicles-functional safety. *International Standard ISO/FDIS*, 2011. 31

[162] Xiaodong Zhang, Matthew Clark, Kudip Rattan, and Jonathan Muse. Controller verification in adaptive learning systems towards trusted autonomy. In *Proc. of the ACM/IEEE 6th International Conference on Cyber-Physical Systems*, pages 31–40, 2015. DOI: 10.1145/2735960.2735971 31

[163] Matthew Clark, Xenofon Koutsoukos, Joseph Porter, Ratnesh Kumar, George Pappas, Oleg Sokolsky, Insup Lee, and Lee Pike. A study on run time assurance for complex cyber physical systems. *Technical Report*, Air Force Research Lab Wright-Patterson AFB, OH, Aerospace Systems DIR, 2013. DOI: 10.21236/ada585474 33, 34

[164] Bojan Cukic. The need for verification and validation techniques for adaptive control system. In *Autonomous Decentralized Systems. Proc. of the 5th International Symposium on*, pages 297–298, IEEE, 2001. DOI: 10.1109/isads.2001.917431

[165] Stephen Jacklin, Johann Schumann, Pramod Gupta, Michael Richard, Kurt Guenther, and Fola Soares. Development of advanced verification and validation procedures and tools for the certification of learning systems in aerospace applications. In *Infotech@ Aerospace*, page 6912, 2005. DOI: 10.2514/6.2005-6912

[166] Chris Wilkinson, Jonathan Lynch, and Raj Bharadwaj. Final report, regulatory considerations for adaptive systems. National Aeronautics and Space Administration, Langley Research Center, 2013. 31

[167] Anh Nguyen, Jason Yosinski, and Jeff Clune. Deep neural networks are easily fooled: High confidence predictions for unrecognizable images. In *Proc. of the IEEE*

Conference on Computer Vision and Pattern Recognition, pages 427–436, 2015. DOI: 10.1109/cvpr.2015.7298640 31, 37

[168] Christian Szegedy, Wojciech Zaremba, Ilya Sutskever, Joan Bruna, Dumitru Erhan, Ian Goodfellow, and Rob Fergus. Intriguing properties of neural networks. *ArXiv Preprint ArXiv:1312.6199*, 2013.

[169] Seyed Mohsen Moosavi Dezfooli, Alhussein Fawzi, and Pascal Frossard. Deepfool: A simple and accurate method to fool deep neural networks. In *Proc. of IEEE Conference on Computer Vision and Pattern Recognition (CVPR)*, number EPFL-CONF-218057, 2016. DOI: 10.1109/cvpr.2016.282

[170] Ian J. Goodfellow, Jonathon Shlens, and Christian Szegedy. Explaining and harnessing adversarial examples. *ArXiv Preprint ArXiv:1412.6572*, 2014.

[171] Nicolas Papernot, Patrick McDaniel, Somesh Jha, Matt Fredrikson, Z. Berkay Celik, and Ananthram Swami. The limitations of deep learning in adversarial settings. In *Security and Privacy (EuroS&P), IEEE European Symposium on*, pages 372–387, 2016. DOI: 10.1109/eurosp.2016.36 31

[172] Vijay D'silva, Daniel Kroening, and Georg Weissenbacher. A survey of automated techniques for formal software verification. *IEEE Transactions on Computer-Aided Design of Integrated Circuits and Systems*, 27(7):1165–1178, 2008. DOI: 10.1109/tcad.2008.923410 32

[173] Mukul R. Prasad, Armin Biere, and Aarti Gupta. A survey of recent advances in sat-based formal verification. *International Journal on Software Tools for Technology Transfer*, 7(2):156–173, 2005. DOI: 10.1007/s10009-004-0183-4

[174] Brian J. Taylor, Marjorie A. Darrah, and Christina D. Moats. Verification and validation of neural networks: A sampling of research in progress. In *Intelligent Computing: Theory and Applications*, vol. 5103, pages 8–17, International Society for Optics and Photonics, 2003. DOI: 10.1117/12.487527 35, 37

[175] Sanjit A. Seshia, Dorsa Sadigh, and S. Shankar Sastry. Towards verified artificial intelligence. *ArXiv Preprint ArXiv:1606.08514*, 2016. 32, 33

[176] Charles Pecheur and Stacy Nelson. V&V of advanced systems at Nasa. *Produced for the Space Launch Initiative 2nd Generation RLV TA-5 IVHM Project*, 2002. 32

[177] Jeannette M. Wing. A specifier's introduction to formal methods. *Computer*, 23(9):8–22, 1990. DOI: 10.1109/2.58215 32

[178] Zeshan Kurd and Tim P. Kelly. Using fuzzy self-organising maps for safety critical systems. *Reliability Engineering and System Safety*, 92(11):1563–1583, 2007. DOI: 10.1016/j.ress.2006.10.005 32

[179] Zeshan Kurd, Tim Kelly, and Jim Austin. Developing artificial neural networks for safety critical systems. *Neural Computing and Applications*, 16(1):11–19, 2007. DOI: 10.1007/s00521-006-0039-9 ix, 32, 41

[180] Luca Pulina and Armando Tacchella. An abstraction-refinement approach to verification of artificial neural networks. In *International Conference on Computer Aided Verification*, pages 243–257, Springer, 2010. DOI: 10.1007/978-3-642-14295-6_24 32

[181] Guy Katz, Clark Barrett, David L. Dill, Kyle Julian, and Mykel J. Kochenderfer. Reluplex: An efficient SMT solver for verifying deep neural networks. In *International Conference on Computer Aided Verification*, pages 97–117, Springer, 2017. DOI: 10.1007/978-3-319-63387-9_5 32

[182] Xiaowei Huang, Marta Kwiatkowska, Sen Wang, and Min Wu. Safety verification of deep neural networks. In *International Conference on Computer Aided Verification*, pages 3–29, Springer, 2017. DOI: 10.1007/978-3-319-63387-9_1 32

[183] Shankar Sastry. Lyapunov stability theory. In *Nonlinear Systems*, pages 182–234, Springer, 1999. DOI: 10.1007/978-1-4757-3108-8_5 33

[184] Insup Lee, Oleg Sokolsky, John Regehr, et al. Statistical runtime checking of probabilistic properties. In *International Workshop on Runtime Verification*, pages 164–175, Springer, 2007. DOI: 10.1007/978-3-540-77395-5_14 34

[185] A. Prasad Sistla and Abhigna R. Srinivas. Monitoring temporal properties of stochastic systems. In *International Workshop on Verification, Model Checking, and Abstract Interpretation*, pages 294–308, Springer, 2008. DOI: 10.1007/978-3-540-78163-9_25

[186] Lars Grunske and Pengcheng Zhang. Monitoring probabilistic properties. In *Proc. of the 7th Joint Meeting of the European Software Engineering Conference and the ACM SIGSOFT Symposium on the Foundations of Software Engineering*, pages 183–192, 2009. DOI: 10.1145/1595696.1595724

[187] A. Prasad Sistla, Miloš Žefran, and Yao Feng. Monitorability of stochastic dynamical systems. In *International Conference on Computer Aided Verification*, pages 720–736, Springer, 2011. DOI: 10.1007/978-3-642-22110-1_58

[188] Zhiwei Wang, Mohamed H. Zaki, and Sofiene Tahar. Statistical runtime verification of analog and mixed signal designs. In *Signals, Circuits and Systems (SCS), 3rd International Conference on*, pages 1–6, IEEE, 2009. DOI: 10.1109/icscs.2009.5412620

[189] Dennis K. Peters and David Lorge Parnas. Requirements-based monitors for real-time systems. *IEEE Transactions on Software Engineering*, 28(2):146–158, 2002. DOI: 10.1145/347324.348874 34

[190] Xiaowan Huang, Justin Seyster, Sean Callanan, Ketan Dixit, Radu Grosu, Scott A. Smolka, Scott D. Stoller, and Erez Zadok. Software monitoring with controllable overhead. *International Journal on Software Tools for Technology Transfer*, 14(3):327–347, 2012. DOI: 10.1007/s10009-010-0184-4 34

[191] Karl Heckemann, Manuel Gesell, Thomas Pfister, Karsten Berns, Klaus Schneider, and Mario Trapp. Safe automotive software. In *International Conference on Knowledge-Based and Intelligent Information and Engineering Systems*, pages 167–176, Springer, 2011. DOI: 10.1007/978-3-642-23866-6_18 34

[192] Philip Koopman and Michael Wagner. Challenges in autonomous vehicle testing and validation. *SAE International Journal of Transportation Safety*, 4(1):15–24, 2016. DOI: 10.4271/2016-01-0128 35

[193] Esther Levin, Naftali Tishby, and Sara A. Solla. A statistical approach to learning and generalization in layered neural networks. *Proc. of the IEEE*, 78(10):1568–1574, 1990. DOI: 10.1016/b978-0-08-094829-4.50020-9 35

[194] Lars Kai Hansen and Peter Salamon. Neural network ensembles. *IEEE Transactions on Pattern Analysis and Machine Intelligence*, 12(10):993–1001, 1990. DOI: 10.1109/34.58871 35

[195] Anders Krogh and Jesper Vedelsby. Neural network ensembles, cross validation, and active learning. In *Advances in Neural Information Processing Systems*, pages 231–238, 1995. 35

[196] Shenkai Gu, Ran Cheng, and Yaochu Jin. Multi-objective ensemble generation. *Wiley Interdisciplinary Reviews: Data Mining and Knowledge Discovery*, 5(5):234–245, 2015. DOI: 10.1002/widm.1158 35

[197] David H. Wolpert. Stacked generalization. *Neural Networks*, 5(2):241–259, 1992. DOI: 10.1016/s0893-6080(05)80023-1 35

[198] Kexin Pei, Yinzhi Cao, Junfeng Yang, and Suman Jana. Deepxplore: Automated white-box testing of deep learning systems. In *Proc. of the 26th Symposium on Operating Systems Principles*, pages 1–18, ACM, 2017. DOI: 10.1145/3132747.3132785 36, 38

[199] Yann LeCun. The mnist database of handwritten digits. http://yann.lecun.com/exdb/mnist/, 1998. 36

[200] Jia Deng, Wei Dong, Richard Socher, Li-Jia Li, Kai Li, and Li Fei-Fei. Imagenet: A large-scale hierarchical image database. In *Computer Vision and Pattern Recognition, (CVPR). IEEE Conference on*, pages 248–255, 2009. DOI: 10.1109/cvpr.2009.5206848 36

[201] Udacity Challenge 2016. Using deep learning to predict steering angles, 2016. https://medium.com/udacity/challenge-2-using-deep-learning-to-predict-steering-angles-f42004a36ff3 36

[202] Contagio 2010. Contagio, pdf malware dump, 2010. http://contagiodump.blogspot.com/ 36

[203] VirusTotal 2004. Virustotal, free service that analyzes suspicious files and urls and facilitates the quick detection of viruses, worms, trojans, and all kinds of malware, 2004. https://www.virustotal.com 36

[204] Daniel Arp, Michael Spreitzenbarth, Malte Hubner, Hugo Gascon, Konrad Rieck, and CERT Siemens. Drebin: Effective and explainable detection of android malware in your pocket. In *NDSS*, vol. 14, pages 23–26, 2014. DOI: 10.14722/ndss.2014.23247 36

[205] Michael Spreitzenbarth, Felix Freiling, Florian Echtler, Thomas Schreck, and Johannes Hoffmann. Mobile-sandbox: Having a deeper look into android applications. In *Proc. of the 28th Annual ACM Symposium on Applied Computing*, pages 1808–1815, 2013. DOI: 10.1145/2480362.2480701 36

[206] Mark W. Craven and Jude W. Shavlik. Visualizing learning and computation in artificial neural networks. *International Journal on Artificial Intelligence Tools*, 1(03):399–425, 1992. DOI: 10.1142/s0218213092000260 37

[207] Jakub Wejchert and Gerald Tesauro. Neural network visualization. In *Advances in Neural Information Processing Systems*, pages 465–472, 1990. 37

[208] Karen Simonyan, Andrea Vedaldi, and Andrew Zisserman. Deep inside convolutional networks: Visualising image classification models and saliency maps. *ArXiv Preprint ArXiv:1312.6034*, 2013. 37

[209] Jason Yosinski, Jeff Clune, Anh Nguyen, Thomas Fuchs, and Hod Lipson. Understanding neural networks through deep visualization. *ArXiv Preprint ArXiv:1506.06579*, 2015. 37

[210] Mariusz Bojarski, Philip Yeres, Anna Choromanska, Krzysztof Choromanski, Bernhard Firner, Lawrence Jackel, and Urs Muller. Explaining how a deep neural network trained with end-to-end learning steers a car. *ArXiv Preprint ArXiv:1704.07911*, 2017. 37

[211] Matthew D. Zeiler and Rob Fergus. Visualizing and understanding convolutional networks. In *European Conference on Computer Vision*, pages 818–833, Springer, 2014. DOI: 10.1007/978-3-319-10590-1_53 38

[212] Shivani Acharya and Vidhi Pandya. Bridge between black box and white box—gray box testing technique. *International Journal of Electronics and Computer Science Engineering*, 2(1):175–185, 2012. 38

[213] Yuchi Tian, Kexin Pei, Suman Jana, and Baishakhi Ray. Deeptest: Automated testing of deep-neural-network-driven autonomous cars. *ArXiv Preprint ArXiv:1708.08559*, 2017. DOI: 10.1145/3180155.3180220 38

[214] Johann Schumann, Pramod Gupta, and Yan Liu. Application of neural networks in high assurance systems: A survey. In *Applications of Neural Networks in High Assurance Systems*, pages 1–19, Springer, 2010. DOI: 10.1007/978-3-642-10690-3_1 39

[215] C. Wilkinson, J. Lynch, R. Bharadwaj, and K. Woodham. Verification of adaptive systems. *Technical Report*, Technical report, FAA, 2013. 39

[216] Kush R. Varshney and Homa Alemzadeh. On the safety of machine learning: Cyber-physical systems, decision sciences, and data products. *Big Data*, 5(3):246–255, 2017. DOI: 10.1089/big.2016.0051 39

[217] Rick Salay, Rodrigo Queiroz, and Krzysztof Czarnecki. An analysis of ISO 26262: Using machine learning safely in automotive software. *ArXiv Preprint ArXiv:1709.02435*, 2017. 40

[218] Fabio Falcini, Giuseppe Lami, and Alessandra Mitidieri Costanza. Deep learning in automotive software. *IEEE Software*, 34(3):56–63, 2017. DOI: 10.1109/ms.2017.79 41

[219] Tim Kelly and Rob Weaver. The goal structuring notation—a safety argument notation. In *Proc. of the Dependable Systems and Networks Workshop on Assurance Cases*, page 6, Citeseer, 2004. 41

Authors' Biographies

SAMPO KUUTTI

Sampo Kuutti received his MEng degree in mechanical engineering in 2017 from University of Surrey, Guildford, U.K., where he is currently pursuing a Ph.D. in automotive engineering with the Connected Autonomous Vehicles Lab within the Centre for Automotive Engineering. His research interests include deep learning applied to autonomous vehicles, functional safety validation, and safety and interpretability in machine learning systems.

SABER FALLAH

Saber Fallah is a Senior Lecturer (Associate Professor) in Vehicle and Mechatronic Systems at the University of Surrey and the Director of Connected Autonomous Vehicle Lab (CAVLAB) within the Centre for Automotive Engineering, where he leads several research activities funded by the UK and European governments (e.g., EPSRC, Innovate UK, H2020) in collaboration with major companies active in autonomous vehicle technologies. His research interests include reinforced deep learning, advanced control, optimization, and estimation and their applications to connected autonomous vehicles.

RICHARD BOWDEN

Richard Bowden is Professor of computer vision and machine learning at the University of Surrey where he leads the Cognitive Vision Group within the Centre for Vision, Speech and Signal Processing. His research centers on the use of computer vision to locate, track, and understand humans. He is an associate editor for the journals *Image and Vision Computing* and *IEEE TPAMI*. In 2013 he was awarded a Royal Society Leverhulme Trust Senior Research Fellowship and is a fellow of the Higher Education Academy, a senior member of the IEEE, and a Fellow of the International Association of Pattern Recognition (IAPR).

PHIL BARBER

Phil Barber was formerly Principal Technical Specialist in Capability Research at Jaguar Land Rover. For over 30 years in the automotive industry he has witnessed the introduction of computer controlled by-wire technology and been part of the debate over the safety issues involved in the implementation of real-time vehicle control.

Printed in the United States
by Baker & Taylor Publisher Services